THE ILLUSTRATED STANDARD OF THE PYGMY GOAT

Written by
Lorrie B. Blackburn, D.V.M.

Illustrated by
Ann Werts

Edited by
Lydia Hale

Published by: All Publishing Company
10951 Meads Avenue, Box G
Orange, CA 92669

Printed in the United States of America
July 1990 Second Printing

Library of Congress Catalog Card Number: 90-082255

Publishers Catalog in Publication Data

Blackburn, Lorrie
The Illustrated Standard of the Pygmy Goat
by Lorrie B. Blackburn, D.V.M.,--1st ed.
p. cm.
Blackburn, Lorrie
I. Goats-Breeds, Pygmy
I. Werts, Ann   II. Hale, Lydia
III. Title
SF383 1986   ISBN 0-9624531-1-0
636.3.9
140p. illus.   9.95 pap

A special thanks to Debbie Edgeworth and Lucy Hulse for their support and encouragement during the development of this book.

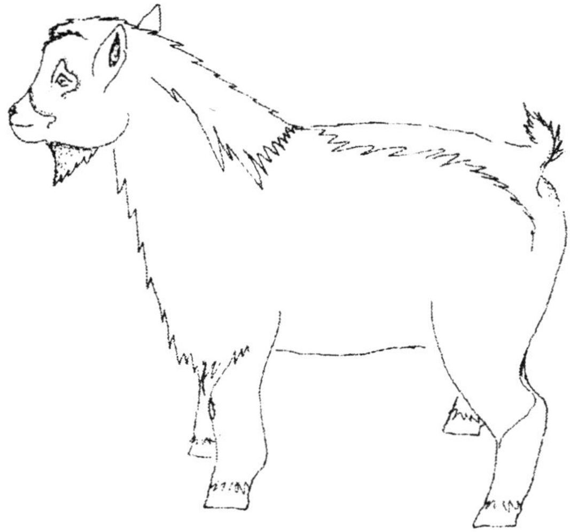

1 2 4 8 16
GOTE

Foreword

Breeders (and judges) of pygmy goats spend many hours observing, comparing and critiquing the conformation of the goats they see. Some breeders pick the best conformation by careful evaluation of many different parts of the goat while other breeders have a "gut instinct" for proper pygmy goat conformation. Both methods can be equally successful for a breeding program. It is the purpose of this book to discuss and explain (in a pictorial fashion) some of the structures and rationale involved in the pygmy goat breed standard. Hopefully this book will help the novice and the expert both to better "see" the conformation features of the pygmy goat and also to understand them.

Good and bad features will be discussed, described and diagrammed. The faults are classified as moderate, serious, very serious, or disqualifying. Judges must disqualify animals with disqualifying faults in a showring. Breeders can be more flexible and must make their own decisions about faults in relation to their particular breeding program. One very serious or disqualifying fault in an animal may be offset by the many good features displayed that are needed in a herd.

The breed standard, Guidelines for Judging, and the scorecard of the National Pygmy Goat Association will be used as the standard for this book. For copies of these documents please refer to the NPGA

Judges Training Handbook or individual documents published by NPGA[a].

[a] National Pygmy Goat Association, PO Box 6037, Bakersfield, CA 93386-6037

TABLE OF CONTENTS

INTRODUCTION

Introduction

The pygmy goat breed standard was originally written to describe the 1975 descendants of the original goats that were imported to this country from Africa. It was describing a healthy, functional animal. A breed standard can exist without change (only occasional clarification) indefinitely. The breed standard is intended to maintain and preserve a breed, not to develop or change one. Natural selection and evolution occur so slowly that those subtle changes won't affect the breed standard for many human generations.

Some parts of the breed standard, such as coloration and general appearance and head, are strictly describing the appearance of the goat. They are specific characteristics of the breed and important parts of the breed standard, but ultimately have no effect on the health or function of the animal. Much more of the breed standard describes the structure of the animal which goes hand in hand with health, longevity, and function.

There are three main categories that are most noticeably effected by structural faults. These are soundness of gait, reproduction, and longevity. Each of these categories is important to the perpetuation of the pygmy goat as a breed.

Conformation of the body proper would be easy to overlook when considering structure that relates to health. The chest and barrel house most of the respiratory, digestive, and reproductive organs of the goat. Breathing and assimilation of food are mandatory for life. Quality life , as opposed to borderline subsistence, demands plenty of room for lungs and stomachs and intestines. A deep heart girth, wide chest floor, great spring of rib and a deep barrel will all help provide the necessary room. A strong topline is necessary to support the soft tissue structures suspended below it. A correct slope to the rump is necessary for ease of delivery of a kid. Each of these body parts is described in its ideal form in the breed standard. And the ideal form is most healthy.

Legs and feet also have a very important role in health versus structure. Legs and feet propel the animal to food and water and they support the animal during breeding. In the buck, the hindlegs are most important. He may be able to function quite well with less than ideal structure of the front legs, but he must use the driving power of his hindlegs when rearing and fighting with other bucks and when mounting and breeding a doe. He must be sound enough on his front legs to willingly pursue a doe in season and to easily travel to food and water, but his rear legs really earn his keep.

A doe, on the other hand, uses her legs differently, but equally as much as a buck. She must travel to food and water. She must support the buck's weight during breeding. And she must carry 15-35 pounds extra weight of uterus and babies during pregnancy. She must have strong enough rear leg structure to lay down and arise without injuring a full udder. And, since 60% of the body weight is carried by the front legs, she must have sound front legs to help support her pregnant body.

Some pygmy goat breeders would disagree with the previous statements about legs and feet. Their management is such that food and water are always available and within easy reach of the animal. The doe in season is brought to the buck and held for him while he services her. A crippled and poorly conformed buck can function under that type of management even if he is quite uncomfortable. He won't function as long as a sound buck, but he will function long enough to produce quite a few offspring. However, since conformation is hereditary, many of those offspring may also become crippled. Management can also nurse a crippled doe through a pregnancy or two, but is that a service to the breed?

Conformation can also have an effect on the successful natural delivery of kids. Conditioning is often not considered as part of conformation but improper conditioning can be very detrimental to an animals reproductive success. Overweight does are difficult to get pregnant. If they do become pregnant, they will have an uncomfortable and stressful pregnancy due to the additional weight of the pregnancy. They tend to exercise much less and therefore are not in good condition for labor and delivery. And finally, a fat doe will have a fat pelvic cavity and will have less space for passage of the kid. An underconditioned doe (unless she is totally emaciated) will conceive and deliver with little trouble, but may have toxic problems due to her lack of proper nutrition. Obese does are also prone to pregnancy toxemia. The overweight buck doesn't have to worry about gestation or delivery, but fat animals tend to be very lazy and less interested in reproduction.

The breed standard describes each of many individual parts of a pygmy goat. All parts of the body must blend together and be proportionate to each other. A random set of perfect body parts will not necessarily produce the ideal conformation. Each part is dependant on others for its proper functioning. Each part must match the breed standard and be in balance with all the other parts of the same goat. Every goat has its faults, and the perfect goat will probably never exist. But we can still strive for perfection and we should keep in mind that the reasons for the breed standard are not just good looks. Longevity is probably one of the most important indicators of good health. We don't want pygmy goats to become like new cars - that fall apart at twelve months just as the warranty expires. Pygmy goats should provide their owners with many years of healthy friendship. We want to perpetuate a healthy and long- lived breed for our descendants to enjoy and much as we do.

PARTS OF THE GOAT

Parts of the Goat

In order to discuss the breed standard of the pygmy goat, it is necessary that you know the parts of the goat and a few basic definitions. The good pygmy goat manager will know, or learn, most of these anatomical parts. It is helpful to know where to look for information, but could become very time-consuming to have to look up most of the information for a conversation about goats.

Throughout this text, frequent mention will be made of "blending." Blending can best be described as the flowing of one section of the body into the next. Blending can be felt far better than it can be seen. As you move your hands over an animal, you should move smoothly and easily from one part to the next. You shouldn't have to change pressure or lift your hand to get around obstacles. There should be no obstacles in a smoothly blended animal. A bucks long, thick haircoat can be very deceptive and can cover more faults than a does shorter, smoother coat. However, smoothness of haircoat should not be confused with smoothness of blending in either a buck or a doe.

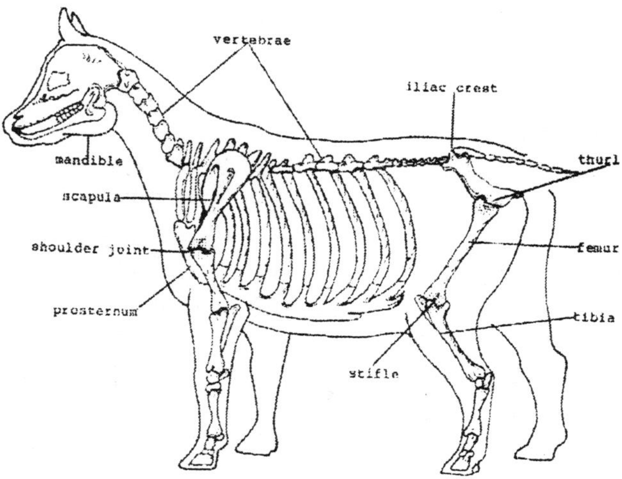
vertebrae
iliac crest
thurl
mandible
scapula
shoulder joint
femur
prosternum
tibia
stifle

Skeleton

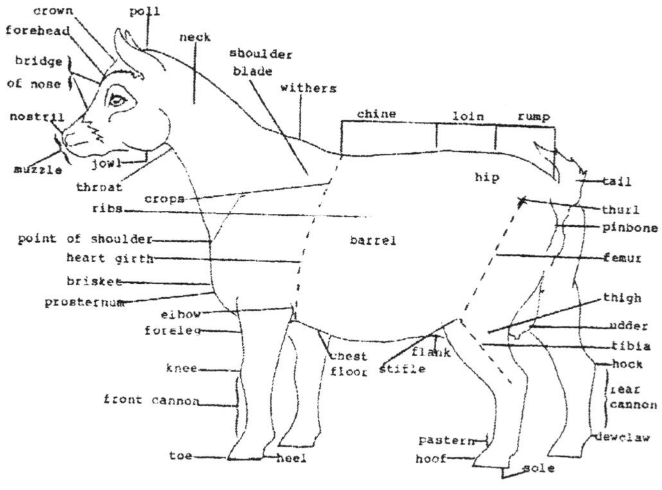

Parts of the Goat

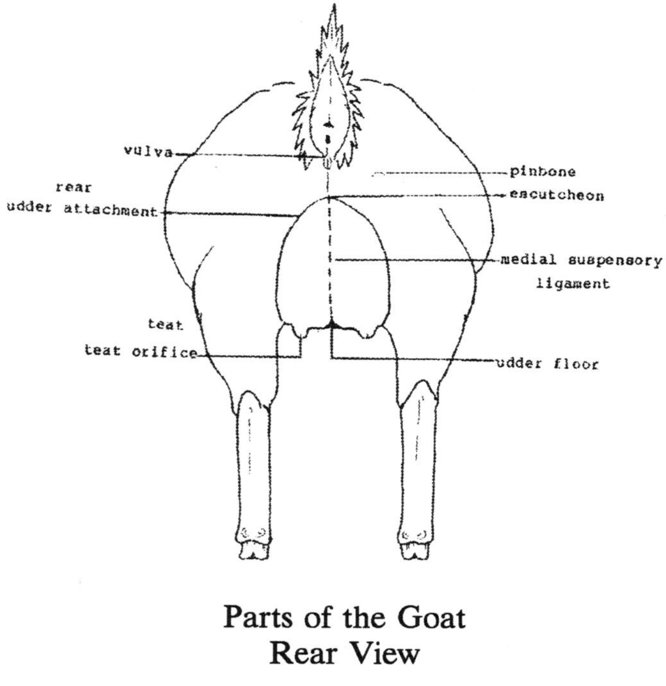

Parts of the Goat
Rear View

GENERAL APPEARANCE

General Appearance
14 points

General appearance is probably the most important category in evaluating the pygmy goat. Many perfect parts and pieces will not produce the perfect goat unless they are correctly proportioned and blend together properly. General appearance accounts for your first impression of a goat as you look at it, and should account for your last impression after you have studied the individual parts separately.

The pygmy goat is cobby. NPGA defines cobby as short legged and thick-set. Compact and square is also used as a definition. The body length, from prosternum to pinbone, is equal to the body height, from withers or rump to ground. A cobby goat has a deep and wide body with short neck, legs, and back. Cobby does not mean the smallest or cutest animal. A 22 inch doe can be as equally cobby as a 16 inch doe. Measurements of withers and cannon are specified for age groups and any questionably short or tall animals should be measured.

The pygmy goat's frame is clearly defined. Management and conditioning of an animal can have a great influence on this aspect of conformation. An overconditioned or overweight animal covers its bone structure with such layers of fat that all body parts blend together. The structure can no longer be seen and often is difficult to feel. Obesity is unhealthy in all animals (and man) and depresses reproduction function. It should be avoided by the careful breeder and manager. An underconditioned animal will have an all too clearly defined skeleton but lacks the smooth blending of parts that is desirable. If an animal is underconditioned to the point of emaciation, it will be disqualified in a showring and obviously should not be used as a healthy breeding animal. Improper conditioning is a moderate to very serious fault.

Health is also of concern as part of general appearance. A judge in a show ring should not try to be a veterinarian and diagnose diseases as part of his placings. However, a conscientious breeder must be aware of general appearance as an indication of general health. Hernias are listed specifically as faults and are usually obvious to any observer. Hernias are abnormal openings in the body structure which allow the protrusion of tissues through them. Common externally visible hernias are umbilical, scrotal, inguinal and abdominal. The first three are often considered hereditary while abdominal hernias are usually the result of trauma. They can all vary in size from minute (1/4 inch) to extremely large (8-12 inches). The judge will be concerned only for general appearance. The breeder should consider the direct effects on reproduction (a large inguinal hernia could entrap a uterus) and possible effects on his future breeding program if the problem keeps cropping up.

Polled goats, naturally hornless, would seem ideal to those breeders who prefer to manage a herd without horns. It would avoid the hassles of disbudding and eliminate any problems with ugly scurs. But polled goats are not allowed registration in NPGA or AGS (American Goat Society). In other breeds of goats, breeding polled pairs produces a high percentage (35) of hermaphrodites and infertile males. Pygmy registries are attempting to eliminate this major source of problems from the breed.

Another important part of the pygmy image is masculinity in bucks and femininity in does. Masculinity is reflected by a majestic carriage, strong neck and shoulders and copious beard and mane. Excess fat or excessive, unbalanced muscling should not be equated with masculinity. Femininity in the doe is shown by a strong but refined neck and shoulders and a well-defined, not overly muscled body. A doe should be so feminine that it is not necessary to check her reproductive anatomy to determine that she is not a wether. The head on both a masculine buck and a feminine doe should excel in pygmy character.

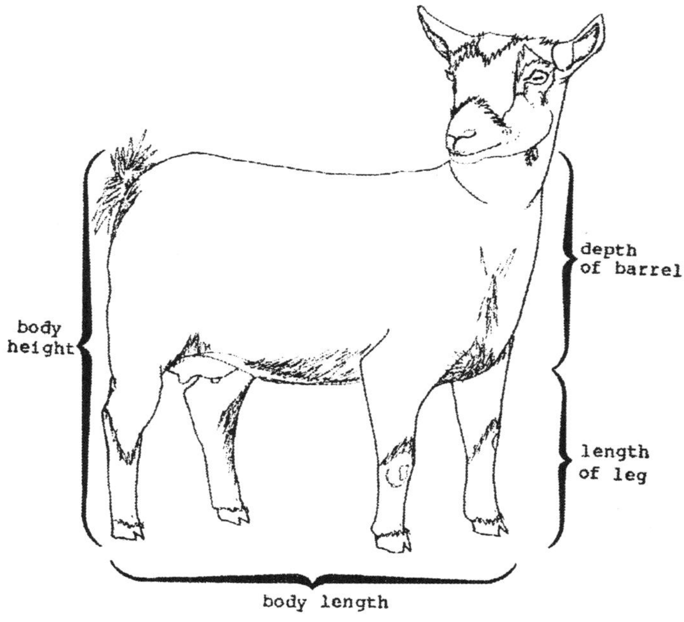

Cobby

The body of a cobby animal should be square. It should also show greater depth of body than length of leg below the level of the barrel.

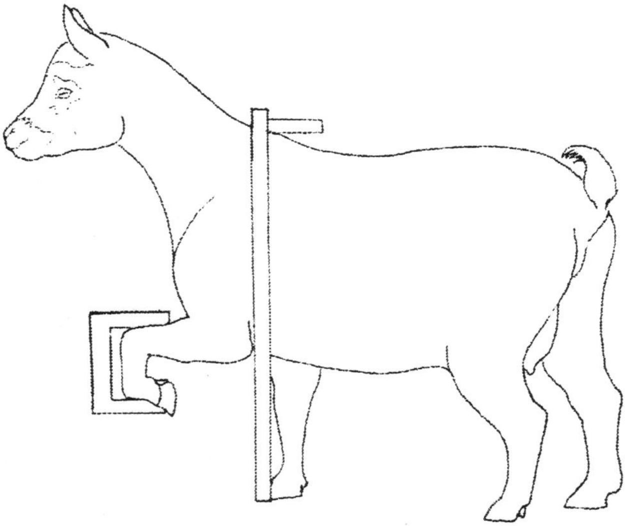

Measurements

Measurements on a pygmy goat should be taken with the animal standing on solid, flat ground. It should be standing squarely with the head held at a natural angle. Holding the head too high will tend to raise the withers and increase the height. Holding the head low will decrease the height measurement. The cannon may be measured on either front leg.

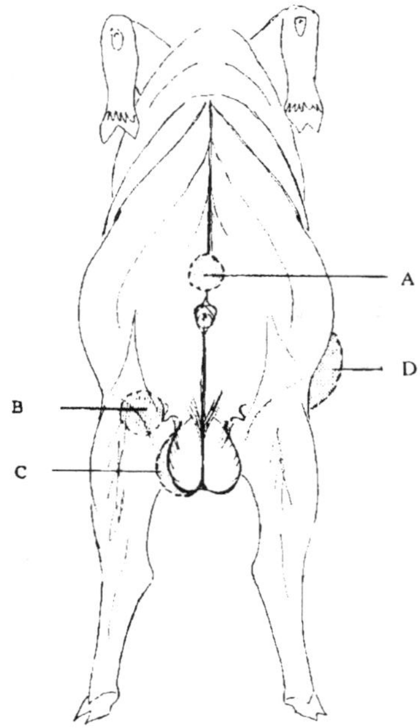

Hernias

Hernias are usually not common in pygmy goats but, if they do occur, they will be seen in four areas. [A] Umbilical hernias are on the midline of the belly at the site of the umbilical cord. [B] Inguinal hernias are high in the groin on either or both sides. [C] Sometimes an inguinal hernia in a buck will spill into the scrotal sac rather than the groin and is then called a scrotal hernia. [D] An abdominal hernia can occur anywhere on the circumference of the abdominal wall.

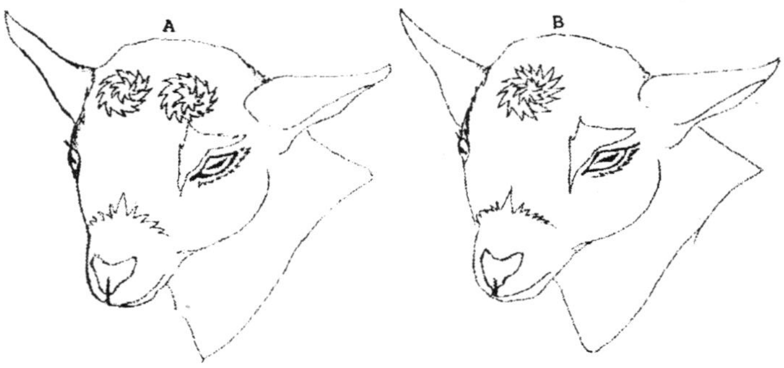

"Hornless" Goats

Naturally hornless (polled) goats are not allowed to be registered as pygmies and are disqualified in a showring. It can be difficult to distinguish between disbudded and polled animals. [A] Any animal that was naturally intended to grow horns will have a swirl of hair centered at each horn location. On disbudded or dehorned animals, a scar can usually be felt at the original horn site. [B] Polled animals will have a single large swirl in the hair of the poll (top of the head). If a polled animal was unnecessarily disbudded, the hair swirls and scars will appear as a non-polled animal.

Scurs

Improper disbudding or dehorning may result in an attempted re-growth of part or all of the horn. These scurs are often small and misshapen but occasionally grow quite large. Some scurs are loosely attached to the skull and are easily broken off. Others have a broad base of bone and require surgical removal. Bucks are more likely to develop scurs than does.

Masculinity

Masculinity

[A] A masculine buck shows proportionately great strength of neck musculature with enough strength of shoulder and chest for smooth blending. The majority of the masculine appearance is the copious hair coat, i.e. mane and beard. Size and mass should not be confused with masculinity. [B] Excess rolls of fat may be deceiving in appearance, but they cannot hide from a careful probing hand.

HEAD and EXPRESSION

Head and Expression
doe-10 points / buck-12 points

Expression is a very subjective feature to be listed as part of a breed standard, but a pygmy with an alert and animated expression is infinitely more pleasing to look at than one that appears depressed or unconcerned with life. Expression is a reflection of head and ear-set and carriage, facial coloration and eye placement. Each of these features has its own place and requirements in conformation evaluation, and their combination and blending create a specific expression for an animal.

The ideal pygmy has a dished face. The dish extends from the forehead (at about the level of the eye) to the muzzle. Animals with horns tend to exhibit a more pronounced dish than disbudded or dehorned animals, but the horns only accentuate the lines of the existing profile. They do not create the dish. Examine the bone structure carefully when evaluating facial profile. Don't be misled by the presence or absence of horns.

Most kids are born with a beautiful pygmy dish. As they mature and the head grows, the profile tends to become more flat or straight, especially in bucks. It is a serious fault if the profile becomes too straight. While we may see many straight faces in our goats, we should always remember to prefer the dished one. A roman nose is the opposite of a dished profile and is a disqualifying fault in a show goat. Do not be confused, however, by the callouses seen on the bridge of the nose of many goats who insist on pushing their noses through wire fences. Constant irritation of the bridge of the nose creates a lump which could be confused with a roman nose unless closely examined. This callous does not extend the length of the face.

The eyes should be set wide apart but should not protrude from the sockets or the side of the head. Protruding eyes are more exposed to injury or infection. Eyes set too far apart - almost on the sides of the head - and eyes set too close together greatly disturb the balance and symmetry of the face. All eye faults are very serious.

The jaws are strong and well muscled and begin their muscling just below the eye. The bone structure of the skull is smooth and the eye sockets are level with surrounding bone, but the added muscle of the jowl makes the jowls wider than the eyes and adds a rounded appearance to the back half of the lower jaw bone.

The ears on a pygmy are medium sized and are held erect. A goat does have a lot of muscular control over it's ears and may actually pull them down to a "helicopter" position in certain circumstances. But if an animal's ears are so large or low set on the head that he cannot hold them erect, then they are a disqualifying fault.

The muzzle should be broad and rounded as viewed from the side, with a wide, flat nose as viewed from the front. A snipey muzzle appears to come to a point from either viewing direction and is very distracting from a beautiful pygmy face.

Bites, or the meeting of upper and lower jaws, are a major area of concern in the pygmy goat. Goats have no front teeth on their top jaw. Instead, they have a very firm structure composed of gum tissue called a dental pad. If this dental pad meets with the lower incisors correctly, the goat has proper occlusion or a good bite. Any deviation from correct is a malocclusion.

Most kids are born with a slightly undershot bite. The lower jaw extends slightly forward of the dental pad. This theoretically helps the kids reach for the teat to nurse. It is only a moderate fault in kids since most will outgrow the condition. Older animals will also tend to be undershot and are faulted more severely. Even with proper occlusion, the lower incisors only meet with a dental pad and they do not wear off very quickly. With age, the gums recede from the incisors and the teeth lean more forward and will reach out past the dental pad. It follows that all middle-aged animals should have a correct bite.

Overshot bites are less commonly seen in pygmy goats. Overshot bites are also called "parrot mouth" since the bite resembles that of a parrot with the upper jaw (or beak) protruding out past the lower jaw. The development of improper bites is not as predictable as one would expect. Some animals with obvious overshot bites at one year of age will have a normal bite by the age of two years. Unfortunately, not all will correct themselves and some animals will continue their malocclusion into their old age. There is also not a predictability of inheritance of bad bites. It is probably safe to breed a nice animal with an overshot bite at least once and watch the kids carefully for two years. They have a relatively good chance of having a normal bite. Overshot bites in older animals will tend to correct themselves with the natural overreach of the lower incisors. Your beautiful two year old animal with the overshot bite may be back in the shoring as a five or six year old! Overshot bites are a very serious fault in kids and adults.

If either an undershot or overshot bite is severe enough to cause the lips to meet incorrectly, it is a disfiguring malocclusion and is a disqualifying fault. Some animals, however, will have a "stiff upper lip" and may appear to have an overshot bite. But once the lips are opened and the bite is actually examined, it will be normal or even undershot. The improper occlusion of the lips and teeth must agree with each other before any improper bite is considered disfiguring.

There seems to be an inordinate amount of interest in the correct occlusion of 8 lower incisors with a dental pad. And we all know that goats are browsers and their incisors are not of big concern. They are not grazers like sheep and do not have to bite grass off close to the ground. But there is a reason for this interest. A goat's bite in the front is a reflection of the occlusion of the premolars and molars in the back of the jaw. It is these larger teeth that masticate the food, chew the cud, etc. that is necessary for proper digestion of food. These rear teeth grow continually throughout the goats life so it is very important that they wear properly on each other. If the bite is misaligned, a front premolar and a back molar will not have a tooth opposing it and will grow abnormally long. Eventually, these long teeth will begin to push on the opposite gums during mastication and will cause pain and the animal will be less willing to chew. The end result will be malnutrition and loss of weight due to malocclusion of the rear teeth.

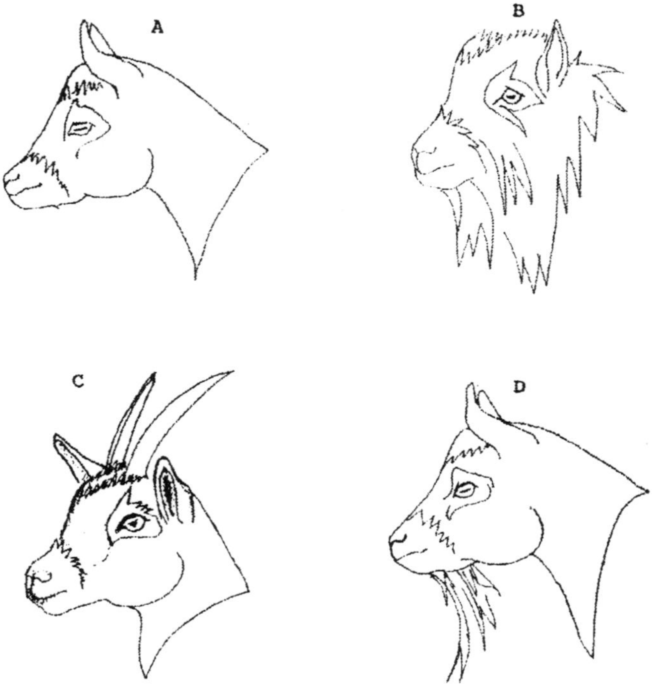

Ideal Head

[A,B] The ideal pygmy head has a dished profile with a wide, broad muzzle and a full, rounded jowl. [C] Horns may accentuate, but will not create, the dish. [D] A beard on a doe is not faulted but may detract from femininity.

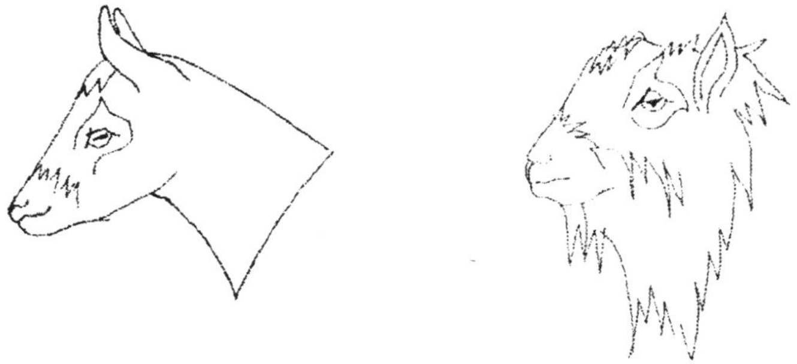

Straight Face

A straight or level profile is not preferred but is common in older animals. Stronger consideration should always be given to mature animals with correctly dished profiles.

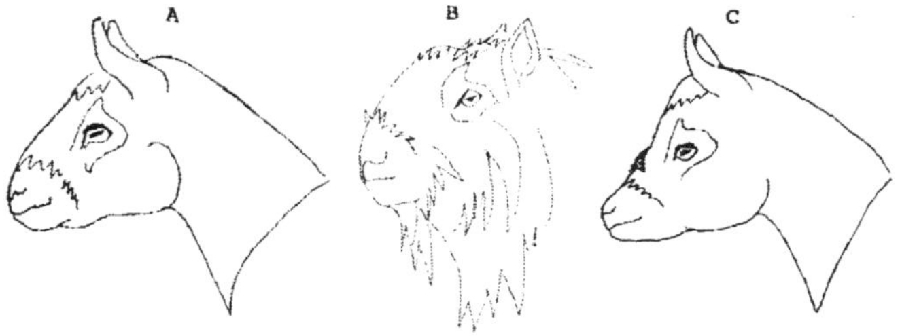

Roman Nose

[A,B] A roman nose is unsightly in does and bucks and directly defies the breed standard. [C] Some animals will have a thickening over the bridge of the nose which, on first glance, may appear as a roman nose.

Ideal Jowls

A pygmy's jaw is strong, well-muscled and adds width to the head. The jowl muscle should wrap around the rear jawbone (mandible) and eliminate the otherwise straight line of the jaw.

Straight Jowls

A straight jawline indicates a relative lack of muscling of the jaw. The muscles of this area are important muscles of mastication and as such are vital to the health and longevity of an animal.

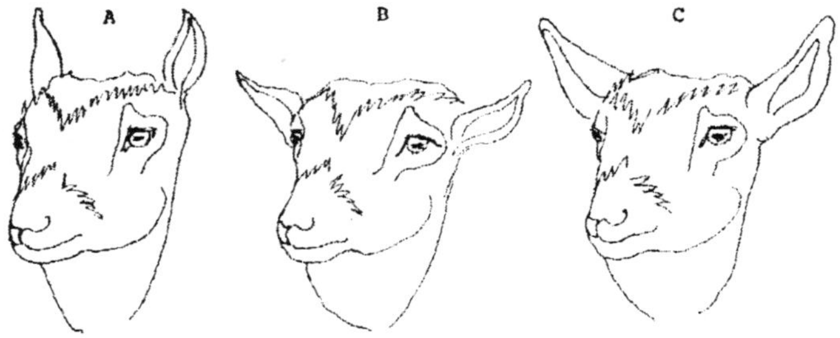

Ears

Ears should be of proper size and placement on the head. If they are set [A] too high or [B] too low, they alter the shape of the entire head. [C] Ears that are too big are more likely to become pendulous.

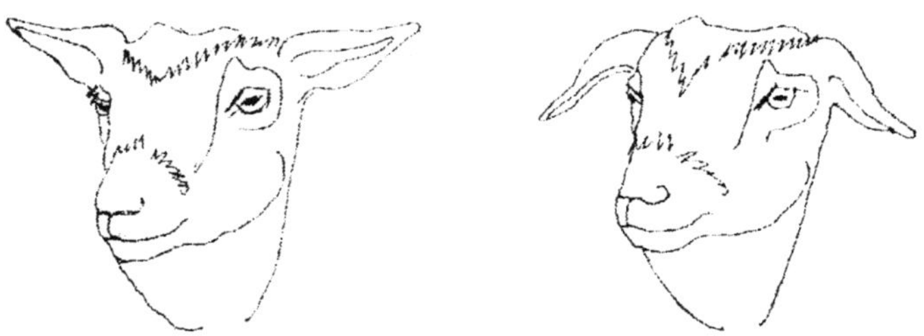

Ears should be erect in the pygmy goat. Either [A] helicopter or [B] pendulous ears are disqualifying faults.

Snipey Muzzle

A snipey muzzle is narrow both from a side view and a front view. It does not add proper balance to an otherwise ideal head.

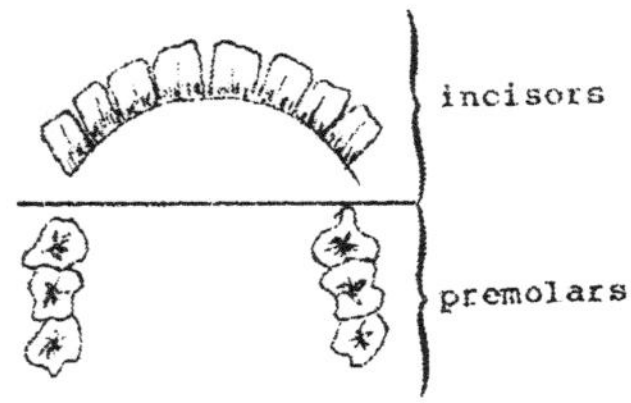

Deciduous Teeth

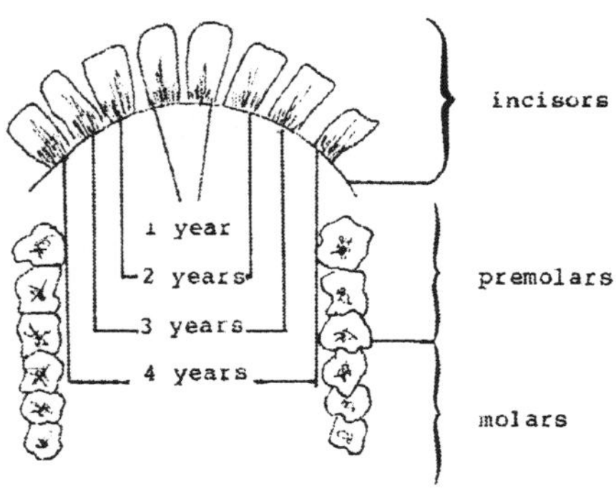

Permanent Teeth

Dentition

Goats have 8 lower incisor teeth, 6 upper and 6 lower premolars and 6 upper and 6 lower molars for a total of 32 adult teeth. The incisors can be used as guides to the age of the animal. Beginning at the middle, at 1 year, a successive pair of baby teeth (deciduous) is replaced by permanent teeth each year. The premolars and molars of the upper jaw match those illustrated in the lower jaw.

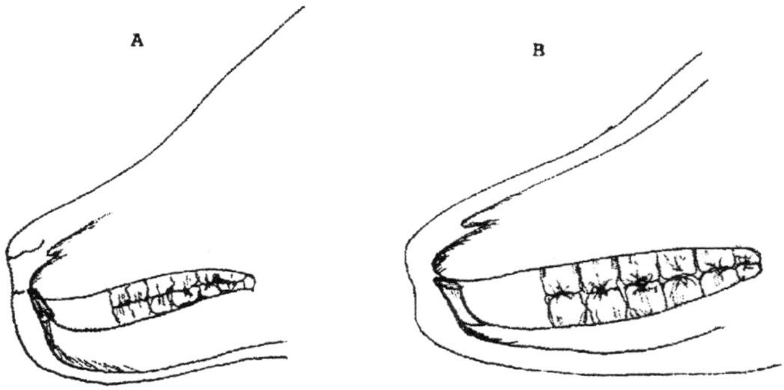

Ideal Bite

The bite, as seen at the front of the goat's mouth, reflects the occlusion of the teeth in the rear jaw. [A] Ideally, the lower incisors meet the lower front edge of the dental pad such that there is no palpable division between the two. [B] The inside edge of the incisor teeth is shaped to fit precisely against the dental pad. This proper fit should extend along all 8 incisors.

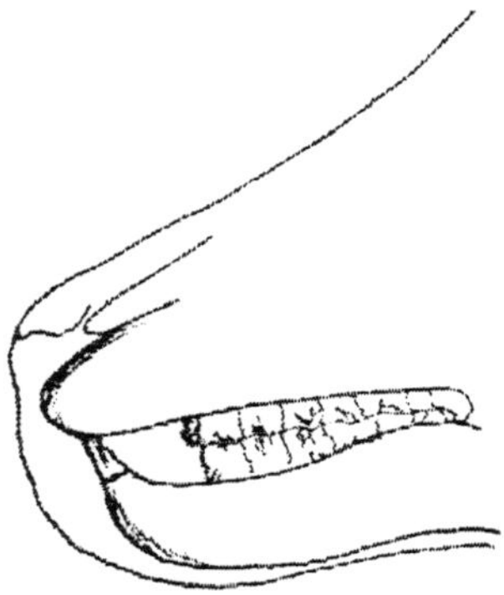

Overshot Bite

With an overshot bite, it is the dental pad which extends in front of the incisors. The incisors push up into the flat of the pad rather than meeting its leading edge. The front upper premolar and the back lower molar will both tend to overgrow and interfere with proper mastication of food.

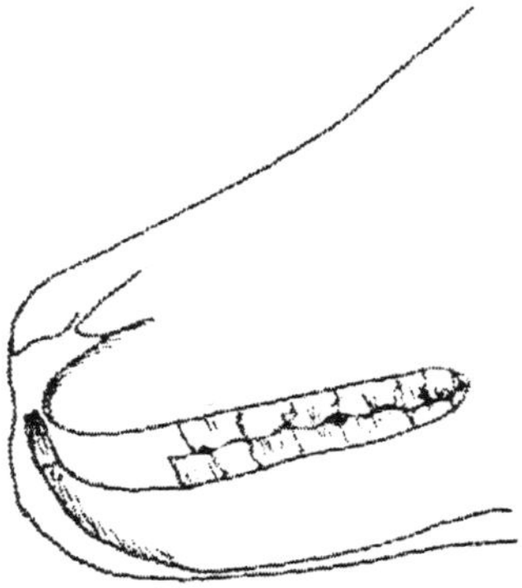

Undershot Bite

With an undershot bite, the incisors extend in front of the dental pad. With more severe undershot bites, the front lower premolar and rear upper molar will not have an opposing tooth to wear against and will tend to overgrow. Undershot bites are common in very young and very old animals.

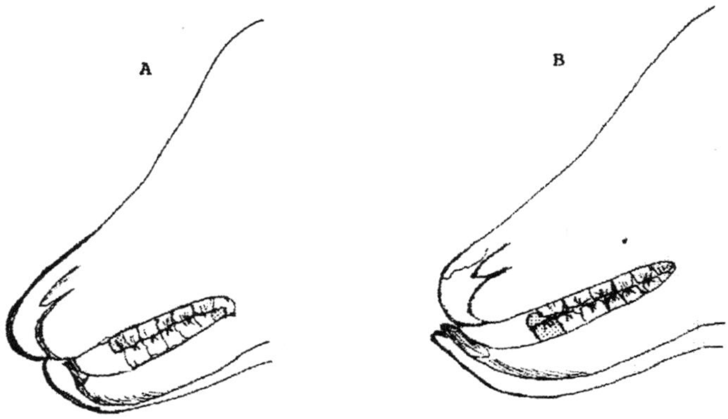

Disfiguring Malocclusions

If an [A] over or [B] undershot bite can be visualized without opening the lips of the goat, it is considered a disfiguring malocclusion. It is not uncommon to have an aged animal show his lower incisors in a "smile", but it is disqualifying in the showring at any age.

COAT and BREED MARKINGS

Coat and Breed Markings
4/6 points   8/12 points

The hair coat in the pygmy is one of the major visible differences between bucks and does. Does have a coat of medium long, straight hair. Some does have a rather fine beard while others have only a few hairs or no beard hairs at all. Many breeders prefer to keep the beards on their does trimmed and this is allowable in the showring. Bucks, on the other hand, have more abundant hair growth over all. Their beard should be long and flowing. They have a long mane that drapes over their neck and shoulders. This haircoat creates a very majestic looking animal.

Curly hair, long hair, and silky hair are all very serious faults. Each of these faults may indicate mixed breeding (i.e. Angora) in the goat's background and are in contradiction with the breed standard, A thin, sparse hair coat, or a coat that is very short and sleek is a moderate fault and may well be a health problem in colder climates.

There are three colors currently recognized in NPGA with a fourth one just beginning to be seen - agouti, black, caramel, and brown. The most "typical" pygmy color is the grey agouti. Agouti is defined in the breed standard as the "pattern produced by the intermingling of light and dark hairs." The grey agouti is the intermingling of black and white hairs to produce a salt and pepper color effect. Depending on the percentage of dark versus light hairs, the colors can range from a light blue-grey to a very dark, almost black agouti. Caramel may range from white to a dark tan. Both brown and brown agouti animals may be seen.

Markings are important in all colors and are the trademark of the pygmy goat. The ears, eyes, muzzle and forehead must have lighter tones (usually white) than the rest of the body color. Solid black goats are the only exception to this rule and they need have no light hairs at all. Darker markings are required below the knees on front legs and below the hocks on the back legs. The dark markings are usually black, but are only required to be darker than the general body color. Brown animals most often show darker brown markings. Caramel animals are allowed a light vertical stripe in their dark stockings on their legs.

If an animal's markings are faded to almost non-existent, it is a very serious fault. Animals may be disqualified in a showring for lacking breed specific markings. The breeder should be aware that breed markings can be affected by their breeding program. Continual breeding of light blue agouti animals will gradually lead to washed out markings. Markings can also be bred back into a herd by careful planning.

Random light markings are totally acceptable if they are in the girth belt area (behind the shoulder and in front of the stifle.) Light patches on shoulders, chest, hindquarters or legs are faults proportional to their distractive value.

Agouti

Agouti

The agouti color is created by the intermingling of light and dark hair. Agouti colored animals must have darker markings (usually solid black) below the knee and below the hock. They most often have a darker mask over part of the face. Lighter or white accented markings are required on the ears, muzzle, forehead and around the eyes.

Caramel

Caramel

The caramel body color can vary from white to a dark tan or brown. These animals are also required to have dark leg markings, but a light vertical stripe is allowed on the front of each cannon. In fact, very few caramel goats will have a solid dark stocking. The facial mask on does will usually be broken and will appear as two vertical lines just to the inside of the eyes. A dark dorsal stripe is also preferred.

Caramel bucks, in addition to the dark markings described for the doe, have a dark martingale over the shoulders. This involves the longer hairs of the mane and creates a very striking appearance.

NECK

Neck
3/5 points

Very few points are given to the neck, but it is a very important part of the body. Not only must it be structured properly but it must be connected smoothly and correctly at both ends.

The neck function is to hold the head away from the body and to provide a means of movement of the head for feeding, observation of surroundings, etc. The neck should be of a short length but balanced to the size of the animal. A neck that is too short can be as displeasing to the eye as one that is too long. It should be well-muscled but not to the point of causing a loss of femininity in a doe or mobility in a buck. The neck must blend smoothly into the shoulders and withers. It should not look like it was stuck on the body as an afterthought. It must also be attached to the body at a correct angle. This attachment accounts for much of the alert pygmy carriage. The head should routinely be carried above the level of the body. An animal who rarely raises its head above that level should be faulted.

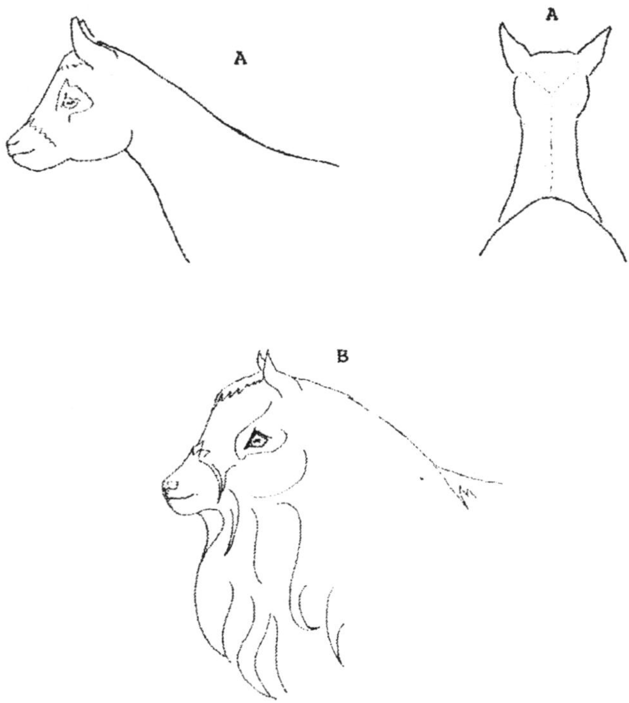

Ideal Neck

The ideal neck should blend smoothly into the jowl and poll and into the shoulders and withers. [A] The neck of the doe is more easily visualized than [B] the neck of the buck which is covered with long hair.

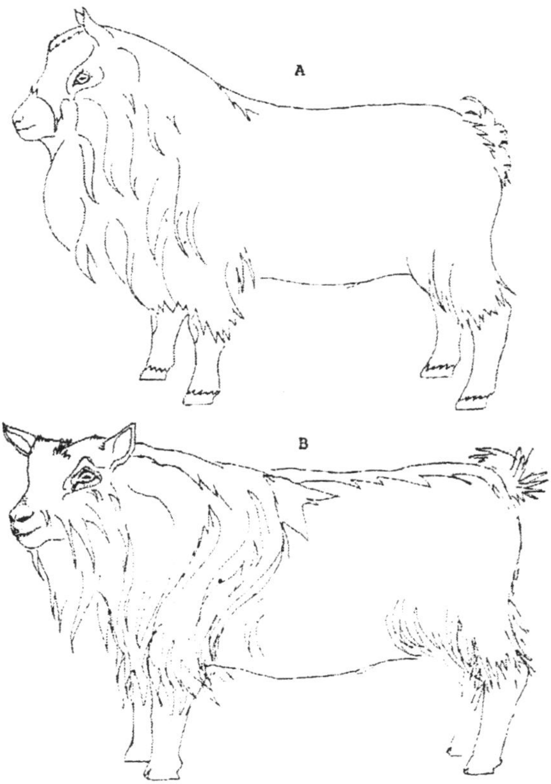

Neck Set

A pygmy goat should be alert. Alert implies an upright head carriage. [A] The angle between the neck and the topline of the goat should be approximately 40° (30°-50°). A greater angle will unbalance the body and [B] a lesser angle causes a loss of alertness and mobility of the head.

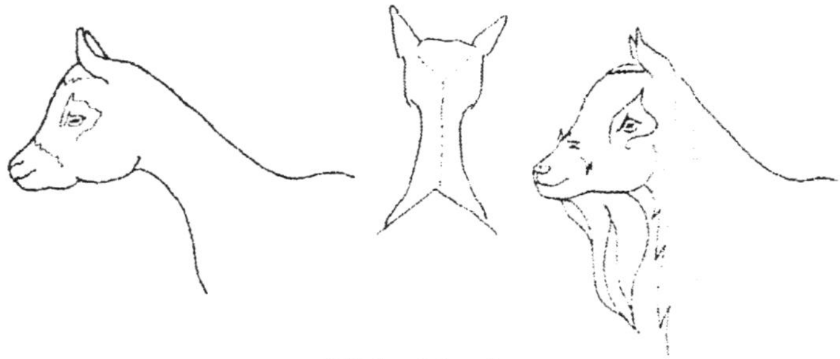

Thin Neck

A thin neck will not blend smoothly into the head or into the shoulders. The goat will appear "top-heavy" with a head that looks too large for the neck that supports it. It will be thin both from top to bottom and from side to side. On a buck, looks will be deceiving and you must feel the muscle of the neck to evaluate its structure.

Ewe-Neck

A ewe-neck dips sharply just in front of the withers. It would almost appear to be set too low on the body, but close examination will show that the dip along the top of the neck is also exhibited along the bottom of the neck. The set into the shoulders and withers is proper, but the carriage is improper.

SHOULDERS

Shoulders
5/5 points

Almost the entire skeleton of the goat is attached by bony joints. But the front leg is attached to the body by soft tissue only. The shoulder is laid on the side of the chest and held there by muscle and ligaments. It is important for these soft tissue attachments to be strong since a lot of weight is hanging between those two front legs.

A well attached shoulder will have the top of the scapula (shoulder blade) just below and to the side of the withers. The withers is actually formed by the dorsal processes of the spinal vertebrae (bony protrusions from the top of the spine).) This dorsal process provides a place for the attachment of the muscles of the shoulder. When examining the anatomy, it is immediately obvious why a perfectly flat withers/ shoulder attachment is not as strong as a correct attachment. Slightly less obvious is the problem with excessive length of muscular attachment and inherent looseness of attachment with too wide or low set shoulders.

The scapula's muscular attachment should be strong enough to prevent the insertion of a persons fingers between the scapula and the body wall. With a loose attachment, this is most easily done behind the shoulder blade near the withers.

An occasional animal will have high withers because of abnormally tall dorsal processes. A quick look may lead you to suspect loose shoulders, but further examination will show that the attachment is strong and the withers are merely high in relation to the neck and chine. This may detract from general appearance on a side view, but will not alter strength or function of the shoulder. Poorly attached shoulders (either loose or high) are a moderate to serious fault depending on the degree of error.

Shoulder angulation is often overlooked in the pygmy goat. The angle found between the humerus and scapula should be about 137°.A greater angle would have the scapula and humerus in almost a straight line and would be very straight angulation. This would put a lot of stress on the shoulder joint and the muscular attachment of the scapula as the animal walks. If the angulation is decreased, the shoulder joint would bear very little weight with all the force being put on the muscles of the scapula attachment and the tissue in front of the shoulder joint. These faults are serious in the showring and in breeding animals.

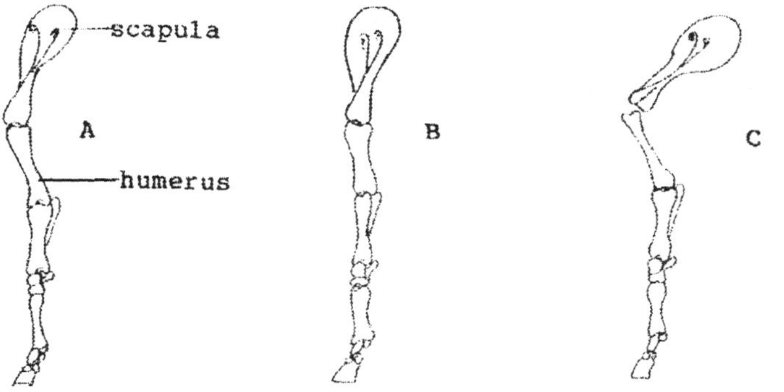

Shoulder Angulation

[A] The correct angle between the humerus and scapula is 137°. This angle will create the right forces on the shoulder joint for health and longevity of that joint. [B] The shoulder joint does act somewhat as a shock absorber for the front leg, and too straight an angle will decrease the shock absorption and will cause stress on the scapula's attachment to the body wall. [C] Too great an angle may cause a tendency for easy dislocation of the shoulder joint. Improper stresses are applied to the joint and the animal will be uncomfortable if required to stand or walk for extended periods of time.

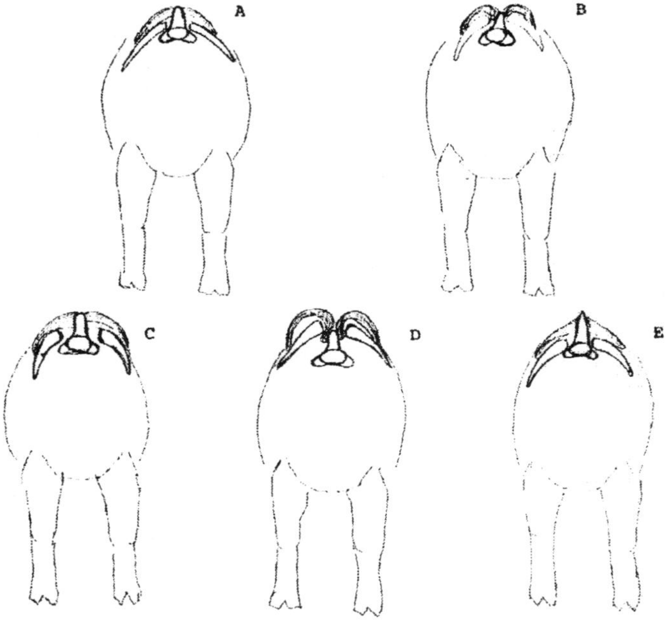

Shoulders

Shoulders

The scapula must be properly aligned with the ribs and vertebrae of the withers to allow the muscular attachments to function correctly. [A] Correct alignment allows a solid, wide base of muscle attachment on the vertebrae and the scapula to hold the shoulder firmly against the body while allowing freedom of motion of the front leg. [B] Level alignment requires that the muscle dip down in between the scapula and the vertebrae to attach. This requires a longer muscle and will cause more frictional wear on the muscle attachment. [C,D] Loose shoulders have very long muscles to bridge the gap. These long muscles allow much more range of motion of the scapula along the rib cage. They will often tend to ride up high over the top of the withers and create much more stress on the entire shoulder attachment. [E] High withers merely create a larger surface for muscle attachment and do not alter the actual alignment of the scapula and vertebrae.

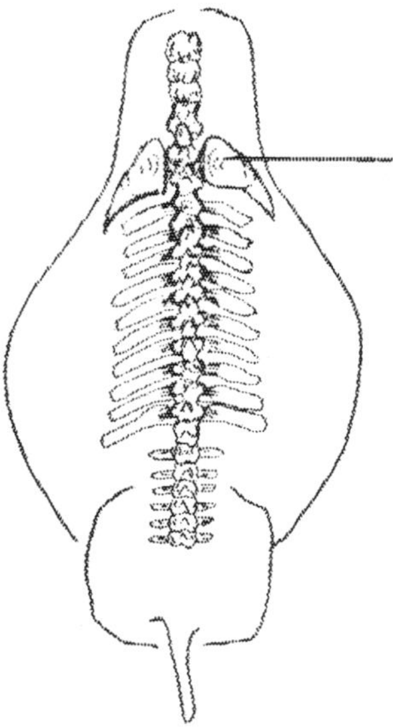

Shoulder Attachment

The scapulas are laid on the side of the rib cage just over the first few ribs. They lay to either side of the dorsal processes of the thoracic vertebrae.

CHEST and BARREL

Chest and Barrel
8/10 points

The chest provides housing for the heart and the lungs and the barrel encompasses all the abdominal organs. These organs need lots of room to properly function. The bony structures surrounding the chest are the spine on top, the sternum below, and the ribs on the sides. The spine and the sternum are each long, narrow bony elements. That leaves the ribs to curve way out to provide the space needed inside the chest. The ribs are attached at bony joints to the spine but they are attached to the sternum by strips of cartilage. This cartilage area tends to be flexible and allows for easy expansion and contraction of the chest as the animal breaths. The barrel's only bony support is the spine.

The floor of the chest should be wide and flat. It is best viewed from the front of the animal and is most often measured between the front legs. The forelegs should be properly set squarely under the body prior to evaluating the chest floor. Improperly set forelegs, too wide apart or too narrow, will falsely alter the true chest floor. A pointy sternum or chest floor indicates that there is less room for heart and lungs in the space above. A chest floor that is overly wide and beefy may indicate a fat content that would also compromise the space available inside the chest.

Viewed from the side, the chest cavity should begin slightly anterior to the shoulders and widen gradually from top to bottom as it is traced back. A prominent prosternum which is visible in front of the shoulder is an indication of a large chest capacity. The heart girth should be deep since a shallow heart girth would be decreasing the chest capacity. A pinched heart girth is a side to side evaluation of the rear part of the chest floor just behind the forelegs and reflects a narrowness where width is desired. The depth from spine to lower body should continue to increase all the way back to the flank.

The chest capacity must also be viewed from above. The chest should widen steadily from front to back. An increasing arch or spring of rib will be evidence of the increased capacity. In does, the increasing width is usually carried through to the barrel and abdomen. But in bucks, there is more likely to be a slight decrease of width directly behind the last rib. This makes it easier to evaluate spring of rib in a buck. Often a doe will have such a large paunch that it will lead you to believe that she has great spring of rib. However, careful examination and palpation of ribs will show the true amount of spring.

It becomes obvious that a slab-sided animal with little or no spring of rib does not have the body capacity for heart, lungs, and intestinal organs as an animal with proper conformation. The same lack of capacity is seen in animals with shallow heart girth and barrels. These are serious faults.

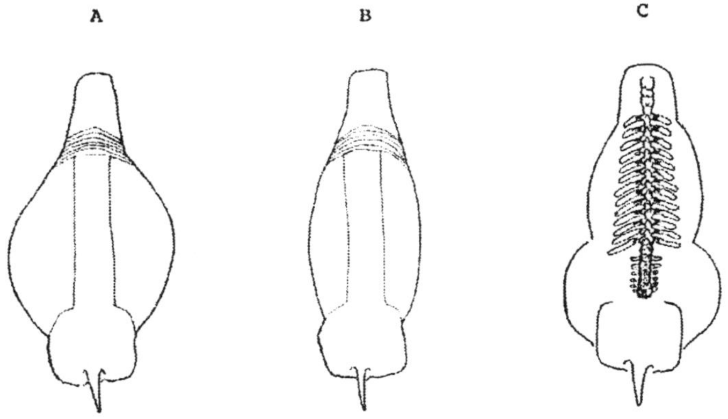

Spring of Ribs

[A] A pygmy should have well-sprung ribs. They should spread progressively wider from front to rear. As they become more sprung, the spacing between each rib will increase. [B] With no spring of rib, the animal appears slab-sided and is lacking in body capacity for internal organs. [C] Some does will have an exceptionally large paunch that will falsely mimic a great spring of rib. Notice should be made of where the last rib is felt in comparison to the width of the spring in that area.

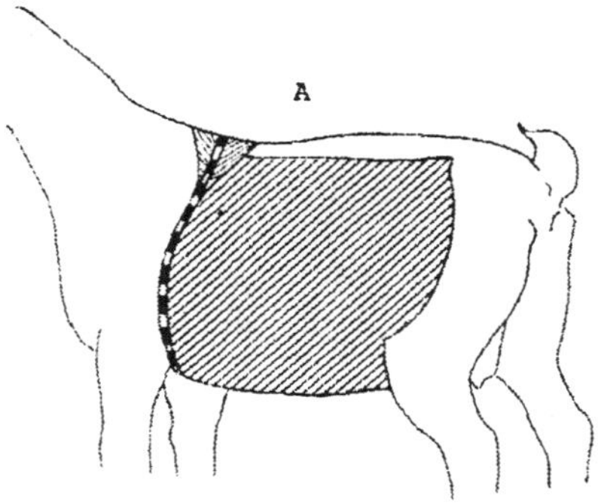

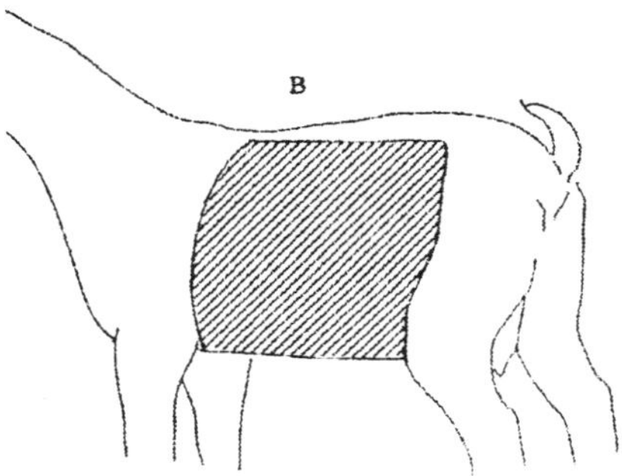

Barrel

The barrel encompasses the body from behind the front legs to just in front of the rear legs. The barrel should also be wide with an increasing width toward the flank.

[A] The barrel should be deep with some increase of depth from front to rear. [B] A shallow animal is lacking in depth of barrel and in body capacity.

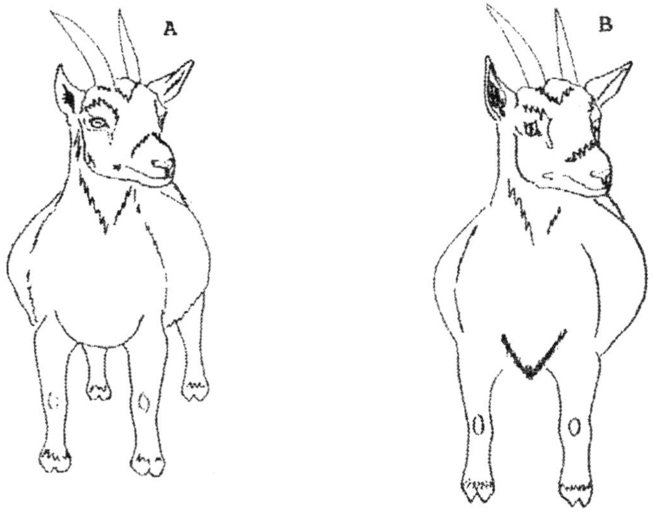

Chest Floor

[A] The chest floor should be wide and strong. [B] Some animals may have good width of chest floor, but will have a lack of muscling on each side of the sternum. These animals will often be pinched in the heart girth.

Lacking Depth of Heart Girth

[A] A pygmy goat should have great depth at the heart girth. The sternum should slope downwards from front to rear. [B] A level sternum will create ay lack of heart girth right behind the elbow.

BACK

Back
8 points

The back is divided into two parts. The chine is immediately behind the withers and ends about halfway to the point of the hip. Anatomically, the chine is the part of the back formed by the posterior thoracic vertebrae (thoracic vertebrae are those with ribs attached to them). The chine should be nearly level. It slopes very slightly down from the withers and then rises very slightly into the loin. Behind the chine is the loin. The loin is formed by the lumbar vertebrae. These vertebrae have wide processes extending from their sides for muscle attachment. So the loin should be wide, well-muscled and strong. It follows the very slight rise of the chine smoothly into the rump.

The most common fault seen in the backs of pygmy goats is a dip in the chine. The withers dip sharply down to the chine and sharply back up to the loin. This dip shows a weakness in the strength of the spine and its associated musculature. A dip in the chine is often accentuated in does in late gestation or in does that are lactating, due to the heavy demands being made on their bodies at those times. If the dip in the chine is followed by a dip in the loin, it is called a swayed back. Swaybacks tend to continually worsen since the weight of the abdominal organs is constantly pulling downward.

A roached back arches upward throughout the chine and loin area. This condition is structurally stronger than a dip or sway, but any deviation from the ideal is weakening to the body.

Backs should be observed while the animal is at rest and on the move. Animals standing still often assume unusual positions. In a showring, the exhibitor is setting the animal up carefully to show off good features and the good exhibitor can hide many faults of the back. The inexperienced exhibitor may accentuate problems with the back. But on the move, the exhibitor has no real control and the back will show its true form.

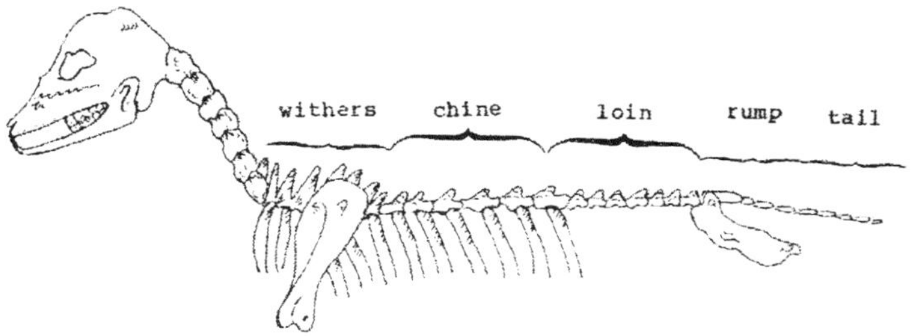

Topline Anatomy

The topline follows the line of the tops of the dorsal processes of the vertebrae. Long muscles travel from front to rear along the sides of the vertebrae to give strength and width to the topline.

Ideal Topline

The ideal topline slopes smoothly and slightly down from the withers into a nearly level chine. The chine blends easily into the loin which maintains a nearly level plane rising only slightly into the rump. The rump begins at the iliac crest (point of the hip) and slopes gently down to a high tail.

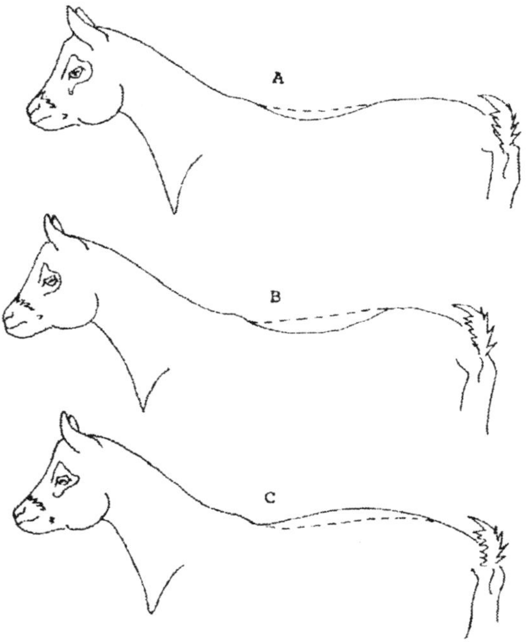

Toplines

[A] A dip in the chine is the most common topline fault seen in pygmy goats. The gentle slopes from withers to chine and back to loin are exaggerated and the topline no longer appears smooth or nearly level. [B] A swayed back dips both in the chine and the loin and is a definite weakness in the structure of the back. [C] A roached back arches upwards through the chine and loin. It is structurally strong but does not give a pleasing appearance to the animal.

RUMP

Rump
8 points

The rump is formed by the pelvic bone. The front end of the pelvis forms the protrusion in the goat called the point of the hip and is located to each side of the midline at the junction of the loin and rump. The back end of the pelvis forms the two pinbones located below and to the sides of the base of the tail. Approximately one-half way between the hips and pins is the true hip joint. This area is called the thurl.

The rump must be observed from two directions. From hips to pins, the rump should slope slightly down toward the tail. It should be of medium length. From thurl to thurl (side to side) the rump should be nearly level with a slight upward arch of the spine in the center. The thurls should be wide apart.

A rump that is either to steep or too level is a serious fault. This is especially important in does since an improperly sloped pelvic cavity can cause difficulty with delivery of kids. A narrow rump and pelvis obviously narrow the space for a kid to pass during delivery.

The width of the pelvis carries through the entire hindquarters. A wide pelvis creates a larger area for udder attachment and a wider base for the hindlegs.

A narrow pelvis does not provide an adequate base for either udder or hindlegs.

The rump must also be observed with the animal on the move. Many animals in a show ring will be scared or angry and will tuck their tail and rump under them. They usually untuck while moving. The experienced exhibitor must try diligently to prevent this problem. It can be difficult for a judge to place an animal as high as it belongs if it constantly shows a steep rump to the spectators.

The tail is set at the end of the rump. With a properly sloped rump, this set will be high,and a happy, healthy animal will carry their tail raised high. The occasional wry tail, which is twisted to one side, detracts from a perfect appearance but usually causes no functional problems. It is only a moderate fault.

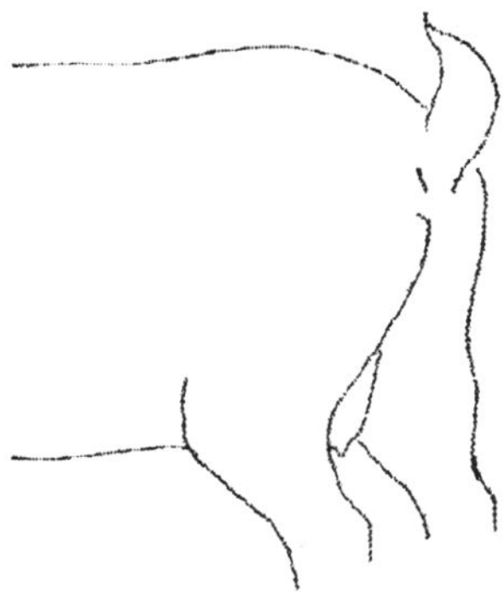

Ideal Rump

The ideal rump in a pygmy has a gentle slope down toward the tail. It should blend smoothly from the loin and into the tail area.

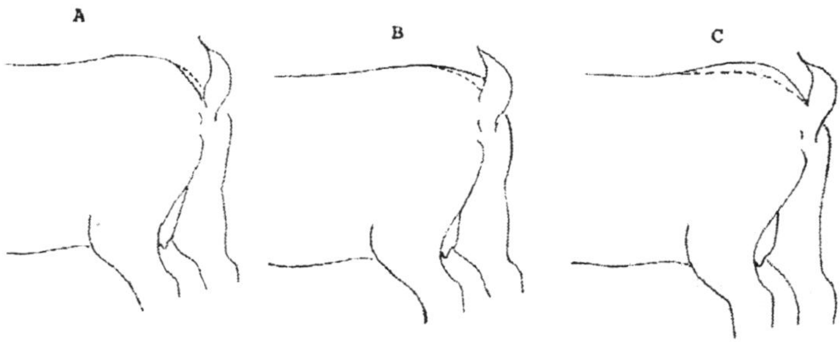

Rumps

[A] A steep rump is not uncommon in pygmies. There is too great a slope from the loin to the tail. This fault often accompanies posty rear legs. [B] A level rump with the loin being at the same level as the tail is rarely seen in the pygmy. Occasionally a buck will have less slope to the rump than is correct but will still not have a truly level rump. [C] A roached rump is most often caused by an unhappy animal that is tucking its hind end under it's body. Since the pelvic bones are solid and will not bend, the roach is a muscular problem.

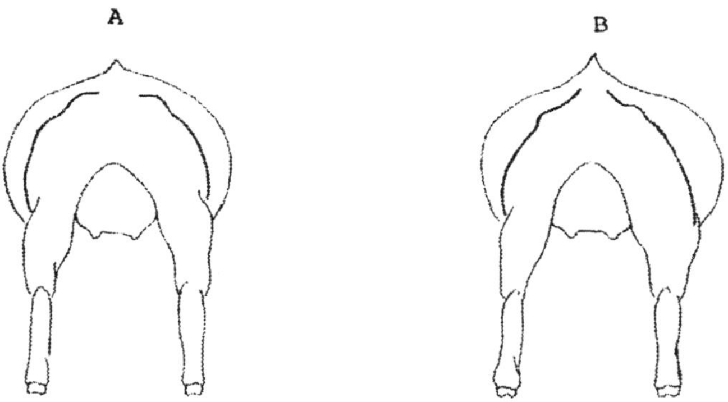

Rump - Rear View

[A] From thurl to thurl, the rump should be nearly level with a slight upward arch of the dorsal processes of the vertebrae. [B] A steep rump, from thurl to thurl, is more common in bucks.

LEGS and FEET

Legs and Feet
10/12 points

The points allotted to legs and feet are very important ones. Most pygmy goats are fed quality foods in special feeders. They do not have to forage all day long in search of adequate nutrition. They still need strong and sound feet and legs to support a strong and sound body. A goat with crippled legs would not last long fending for itself in the wild. It might also have trouble travelling from feeder to waterer and back in a pygmy goat fancier's goat yard.

Legs also need to be viewed from two directions - front and side for front legs and rear and side for rear legs. The front view of the front legs should show strong straight legs set wide apart. They should be set squarely on the side of the chest above them. The elbows should point directly backwards and should be carried in close to the ribs. A plumb line dropped from the point of the shoulder should pass through the elbow, knee, and between the cloves of the hoof. Any deviation from this straight line puts abnormal stresses on joints and muscles and is a serious fault.

Kneepads are not part of the breed standard, but they are present on almost all sound goats. Goats use their knees to support their weight both when lying down and when getting up. A callous develops quite naturally as a result of this constant wear. An occasional goat will not have calloused kneepads. The most common reason would be a front leg lameness which causes the animal to protect its knees or upper arm by not bearing much weight on the bent knee. Another possible reason would be an animal kept on exceptionally soft ground and/or bedding which cushions the knee adequately to prevent callous formation even with normal wear. It behooves a judge to watch carefully for any abnormalities of gait in an animal without kneepads. And it would be very good management for a breeder to be especially suspicious of those animals.

Viewing the front legs from the side should also show a straight leg from elbows to pastern. The pasterns are short and strong and slope slightly forward. The pasterns act as a shock absorber when the animal moves.

Buck knees and knees that bow backwards are both serious faults and cause abnormal stress on the knee joint. Pasterns that are too long or sloped are weak and tend to put extra stress on the flexor muscles of the back side of the leg. Very straight pasterns lose their shock absorbing ability.

Moving back to a side view of the rear legs, correct angulation becomes important. It is actually hock angulation that is discussed. This is most easily seen by another plumb line. Drop the line from the pinbone to the ground. It should pass just along the back edge of the hock and cannon.

The most common angulation fault in the pygmy goat is postiness. The hock angle is too straight and the vertical cannon sets way in front of the plumb line. Animals with this fault tend to be stiff in the hind leg motion and often swing their legs out to the side of the body to travel forward rather than flex the joints normally. Sickle hock is less common and is the opposite of postiness. The hock is too angulated. If the cannon is set vertically, it would be way behind the plumb line. This puts too much strain on the muscles to hold the leg in its normal position.

The last view of the legs is to examine the rear legs from the rear. The legs should be straight from wide set thurls. The stifles should point forward and the hocks should point directly backwards. As with structure of the front legs, any deviation from straight conformation is a serious fault.

Hocks that turn in (cow-hocked) are a common fault. Many pregnant does will be cow-hocked due to pressure of the expanded abdomen pressing outward on the stifles, which causes the hocks to turn inward. There is no similar excuse in bucks or in young does.

(As an aside, I must question the correctness of the

breed standard in requiring straight hindlegs when viewed from the rear. Examination of San Clemente wild goats and deer and other "goat-like" animals in the wild reveals a fairly marked degree of cow-hockedness. These animals are very agile and successful breeds without man's influence. Will our requirements for straight rearlegs eventually cause ambulation problems in our goats? Or will enough animals show fine pygmy conformation except for being cow-hocked that we will retain soundness?)

All four feet should be well-shaped and symmetrical. The two cloves of the hoof should be parallel to each other with only a slight space in between them. The heels should be deep with the coronary band running parallel to the ground.

Splayed feet are a very common fault in pygmy goats, especially in certain areas of the country. Some splayed feet merely separate more at the toes than at the heels. Other splayed feet have extra growth of hoof wall between the two cloves which forces them apart. Badly splayed feet may eventually cripple an animal. When examining for splayed feet, be sure to check at least one front and one rear foot. One may be quite differently formed than the other. Careful trimming may help prevent progress of splayed feet, but will usually not avert the condition entirely.

Crooked hooves may also be observed in the pygmy

goat. Below the coronary band, the hoof will bend inwards (usually) or outwards and one edge of the hoof wall will become the weight bearing surface of the foot. Again, hoof trimming will help prevent progression but won't prevent or cure the problem entirely.

Gait, the way of moving, must also be considered under feet and legs. Beautiful looking legs are useless if they cannot propel the animal properly. Ideally, correct structural conformation will go hand in hand with smooth, flowing movement. Many structural faults may be associated with specific problems with gait. In a proper gait, an animal tracks with its feet squarely below its body and moving forward in a straight line. Legs that swing in or swing out are indications of a problem. Movement should be easy, not choppy and bobbing.

All the structural faults of the legs and feet are serious. Faults of movement are very serious. One might be wise to be extremely critical of rear leg conformation in bucks since rear legs are so important in his performance of reproduction. Without strong functional rear legs, he will never have to opportunity to pass on the rest of his good characteristics to any offspring.

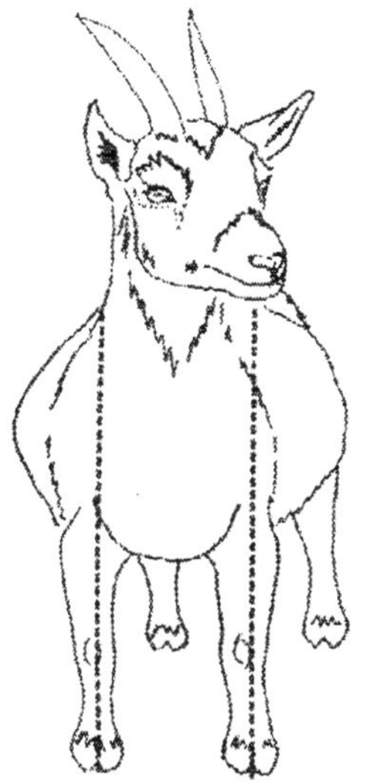

Front Legs - Ideal Set

The front legs should be strong and set wide apart. A plumb line dropped from the shoulder joint to the ground should pass through the center of the elbow, the knee, and the hooves.

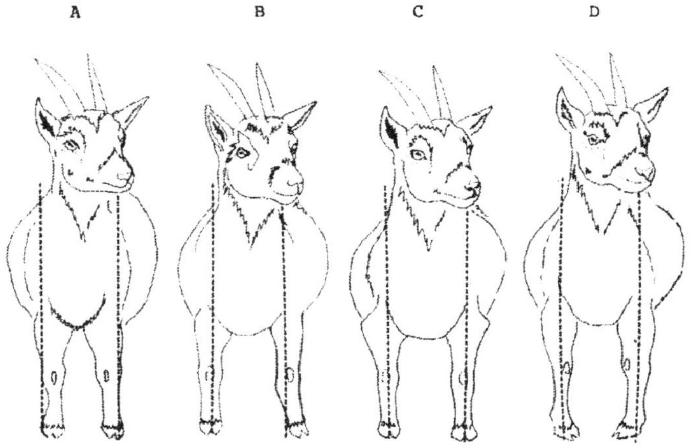

Front Legs

[A] Front legs that are set close together will accompany a narrow chest floor. It is important to see that the legs are set straight vertically before evaluating how close or wide they are carried. [B] Crooked legs are more often seen in younger kids and will often correct themselves as the animal grows. The leg will be straight from shoulder to knee and then veer out away from the plumb line. [C] Some animals will carry their front legs with the elbows pulled out away from the chest wall. This will not give good support to the movement of the fore quarters. [D] Animals that toe in will often also be out at the elbows. This will cause abnormalities in gait.

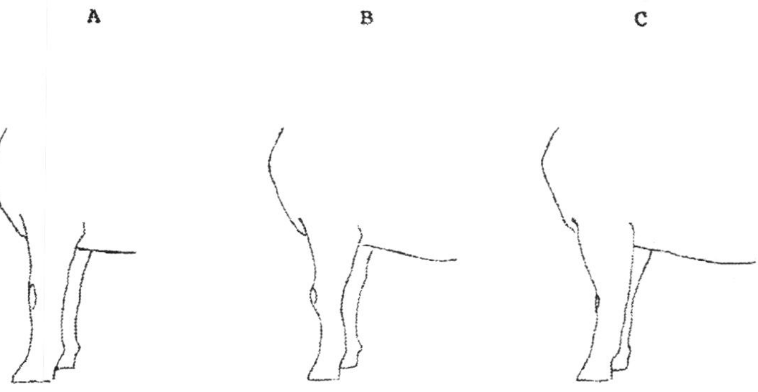

Front Legs - Side View

[A] The ideal front leg is straight from elbow to pastern. [B] Buck-knees bend forward at the knee and do not fully straighten even on the move. [C] An occasional knee will bend slightly backwards. Muscles are stretched or torn and most all animals with this fault will be lame.

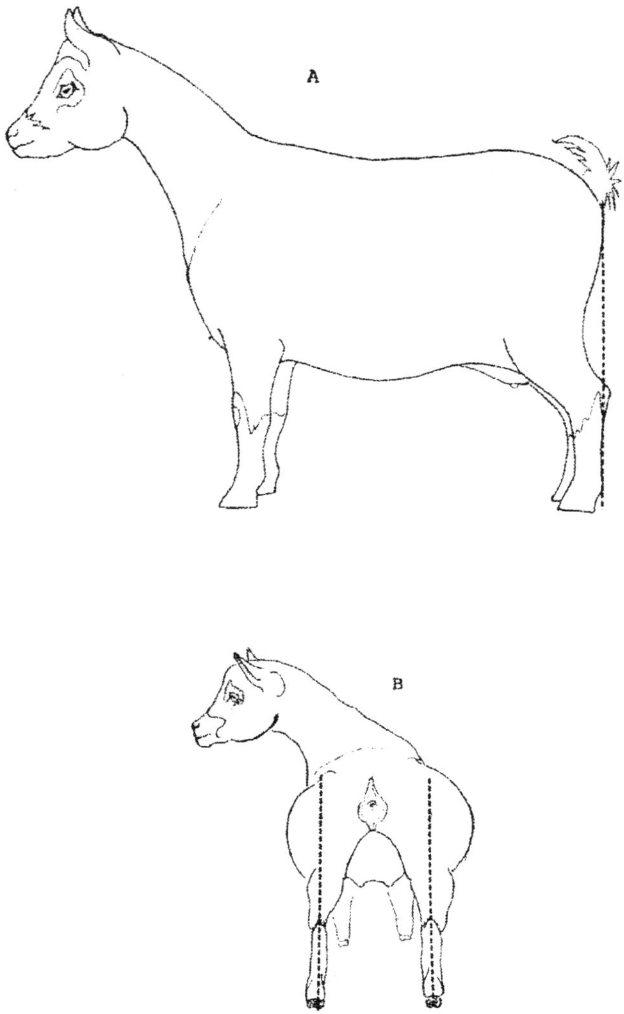

Rear Legs - Ideal Set

Rear Legs - Ideal Set

[A] The side view of the rear legs should show well angulated hocks and a pronounced stifle joint. A plumb line dropped from the pin bone should fall along the back of the cannon. [B] From the rear, the hindlegs should be widely set on a wide pelvis. The plumb line is dropped from the thurls and should pass through the stifle, hock and cannon.

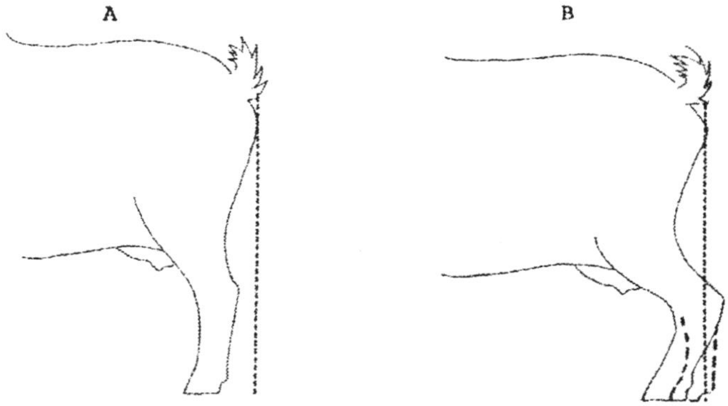

Rear Legs - Side View

[A] Posty legs are a very common fault in pygmy goats. The hock is almost straight and if the leg is set with the cannon in a vertical position, the plumb line will fall way behind the cannon bone. All the muscles are shortened and will cause alterations in the animal's gait. [B] Sickle hocks are a less common fault. The hock is excessively angulated and if the cannon is set vertically, the plumb line will fall in front of the cannon bone.

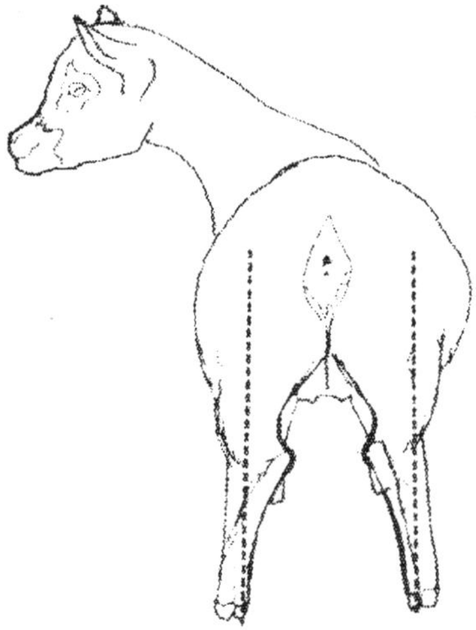

Rear Legs - Rear View

Many pygmy goats are cow-hocked. The hocks turn in toward the center of the body and the stifles necessarily point slightly away from the body. This condition is especially common in does in late gestation.

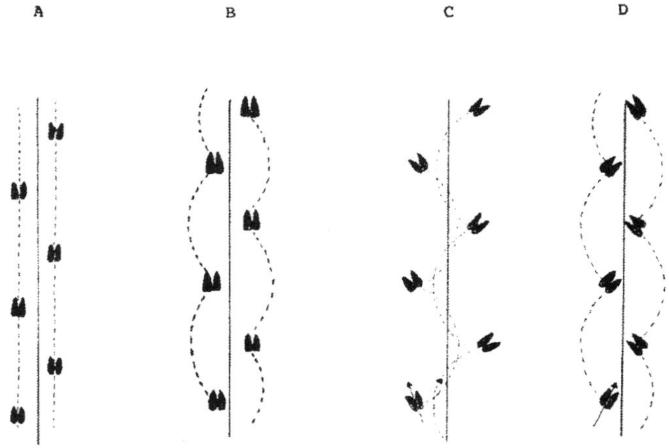

Tracking

Tracking

Observation of gait from the rear of the animal will allow evaluation of the way in which the animal tracks with his rear legs. [A] Ideally, the leg is bent and the foot lifted from the ground. The leg is swung forward in a straight line and the foot replaced on the ground. [B] Many posty legged animals cannot adequately stretch their shortened muscles to allow complete flexion of the joints of their rear legs. They lift the foot from the ground by shifting their weight to the opposite hindleg and pulling the leg and foot out to the side. Then they swing it out and around to a position in front of where it started. [C] Cow-hocked animals tend to swing their legs inward as they move forward.

The same tracking maneuvers apply to the front legs. [B] Animals that toe in for any reason will swing their forelegs out as they move. This is called paddling. [C] Animals with crooked legs or those which toe out will swing their forelegs inward.

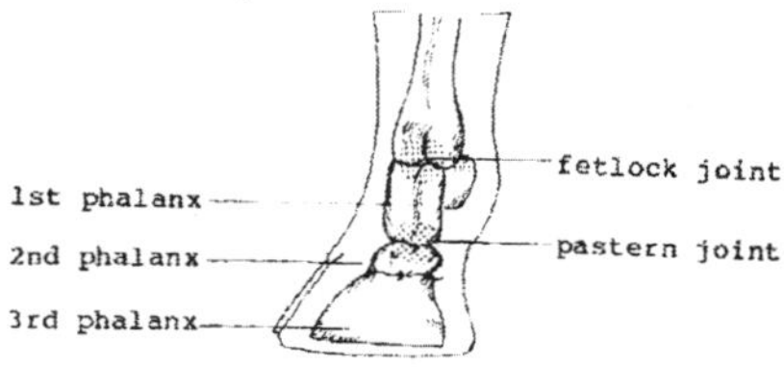

Pastern Skeleton

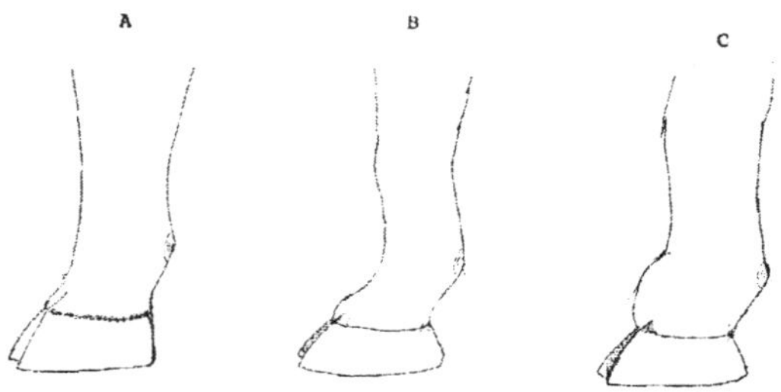

Pasterns

[A] The ideal pastern is short and strong. When properly angulated, it is resilient and an excellent shock absorber for the front leg. [B] A weak pastern is over angulated and puts extra stress on the muscles along the back of the leg which keep the pastern from bending to the ground. Some animals with weak pasterns will eventually break down to the point where they do walk on the pasterns and dewclaws. [C] A bowed pastern is caused by incorrect angulation of the pastern joint. The stresses on the leg muscles are not changed, but the joint will quickly become arthritic and painful to the animal.

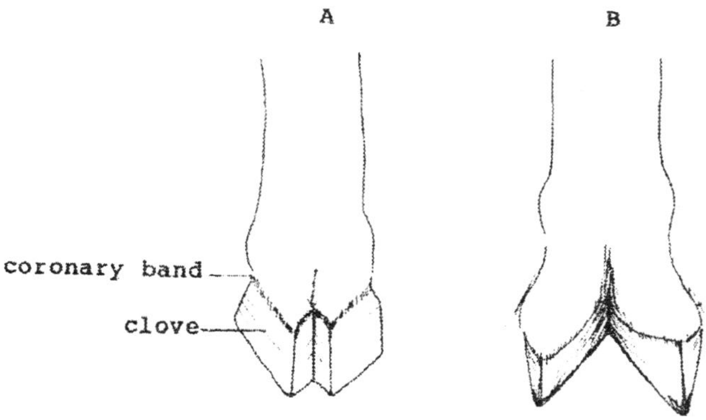

Hooves - Front View

[A] Ideally, the cloves of the hoof should point forward and be close together. [B] In splayed feet, the tips of the toes point away from the midline. Some feet will be splayed so severely that the toes almost point in opposite directions.

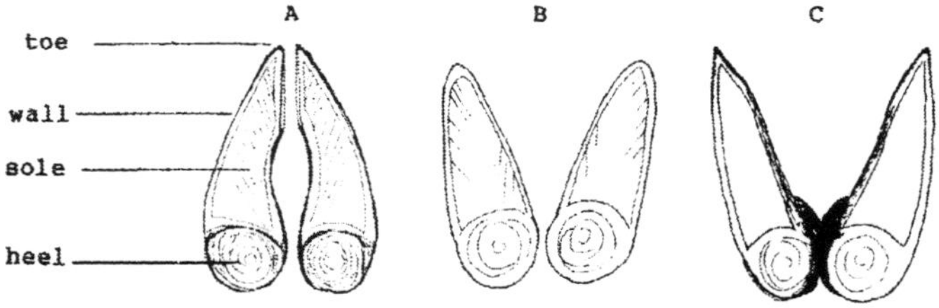

Hooves - Bottom View

[A] In a properly trimmed hoof, the heels and toes should be equally close to each other. [B] A splayed foot has the heels close together but the toes are separated. [C] A severely splayed foot will have an overgrowth of wall along the inside rear edge of each clove. This overgrowth can become quite severe and may extend well up towards the coronary band.

BUCK - REPRODUCTIVE AND MAMMARY SYSTEM

Buck
Reproductive and Mammary Systems
0 points

No points are allotted to the reproductive or mammary system of a buck. This is an all-or-nothing situation. If he is formed incorrectly, he is disqualified from showing and most often will not be used much, or at all, in a breeding herd. If he is built correctly, then he can stay in a showring and compete and will be more welcome in the breeding pens with the does.

A buck must have two testicles, of approximately equal size, normal texture (you have to feel several bucks to decide what normal is!), and fully descended into the scrotal sac. A buck with one or both testicles missing from the scrotum is called a cryptorchid. The missing testicle is hidden in the abdomen or inguinal ring - never having completed its proper descent to the scrotum. An occasional buck may have a tumor or mass in a testicle causing it to be larger than the normal testicle. Others may have a hormone or injury-related problem causing one testicle to be abnormally small. Some of these problems, especially cryptorchidism, are considered by many geneticists to be hereditary, and we should be very careful not to breed such problems into the pygmy goat.

Bucks do have two small, non-functional teats just in front of the scrotum. Any supernumerary or double teat or a teat with multiple orifices is disqualifying in a buck. A functional teat which produces milk is also a disqualifying fault. However, remember that bucks get very oily and wrinkled skin in their groin area and sometimes the extra folds of skin may appear as an extra teat. Careful examination usually eliminates any doubt. If doubt remains about a "suspicious lump", a judge in the showring should give the benefit of the doubt to the animal and assume a normal mammary system. A breeder may wish to be more stringent in his/her evaluation.

It is very helpful, even mandatory, to use your hands to evaluate a bucks reproductive and mammary system. Testicles must be felt for consistency and size. Feeling the teat and the area immediately surrounding it may reveal the scar from a previously removed supernumerary teat. Some extra teats are pigmented so they are hard to see, but color can't distract your sense of touch. While you're touching, tip each teat so you can examine its orifices and assure yourself that there is only one.

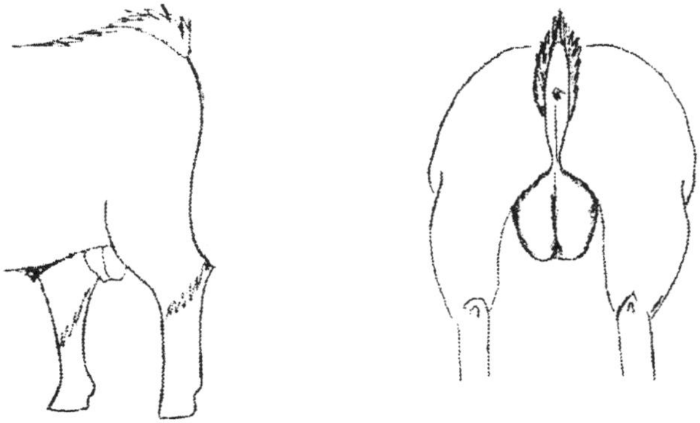

Correct Buck Reproductive System

There is not an ideal conformation for the buck's reproductive system. There is only correct and incorrect. A buck must have two testicles fully descended into the scrotal sac. The testicles should be equal in size and evenly descended.

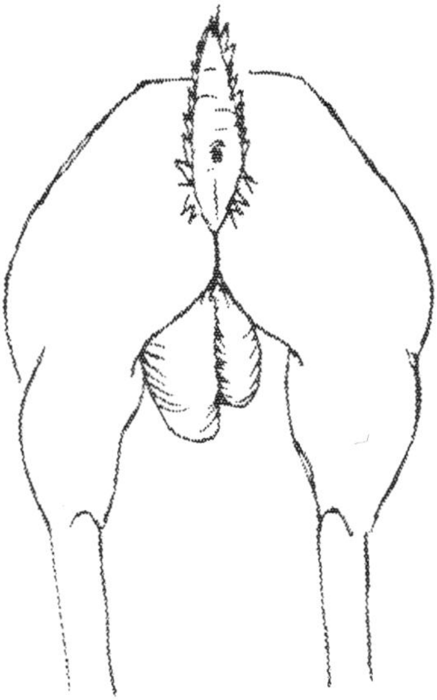

Unevenly Descended Testicles

There is no fault for unevenly descended testicles unless one is held way up out of the scrotal sac. But an uneven scrotum should be carefully checked for equal size and texture of testicles.

DOE-MAMMARY SYSTEM

Doe - Mammary System
12 points

The mammary system of the doe is the udder. Included as parts of the udder are capacity, shape, front and rear attachments, texture, and teats. Since 12 points are allotted for the entire mammary system, that allows 2-3 points for each part. Capacity, shape and attachments can only be properly evaluated in a lactating animal with a full udder. Does are only occasionally shown with properly bagged udders, so that leaves the majority of udder evaluation to be done by the breeder in his/her back yard.

The pygmy goat udder should have the capacity to feed all the kids produced by that doe. Few does will ever be able to feed quads or quints without some supplementation by the breeder. But does should routinely be able to support twins or triplets. The full udder should have a pleasing globular shape and be evenly balanced on both sides. (A goat udder has two halves, side by side, which function separately from each other.) The full udder should hang between the hindlegs with a portion showing behind the legs and a portion showing in front. It should not be so large that it interferes with locomotion.

An udder that is larger on one side than the other is unbalanced. This is most often due to single kids nursing only on one side but may also be due to a tight teat sphincter that is difficult to nurse, or mastitis which causes pain to the doe so she won't allow kids to nurse on one side. Regardless of the reason, an unbalanced udder is a serious fault in a showring.

The udder attachments are important to the longevity of a healthy udder. The front attachment should blend smoothly into the belly of the doe. There should not be a "pocket" or recess right in the middle at the front edge of the udder. The side attachments should be wide and seem to run right into the inside of the legs in the groin area. The rear attachment is the one attachment that can be somewhat estimated in young or dry animals. The rear udder attaches at the escutcheon. A high and wide escutcheon means a high and wide rear attachment. An udder with strong attachments all the way around will be held tight to the body wall along its top surface. When the attachments are palpated by trying to encircle the base of the udder with your hands, you would hope for a wider base of attachment than the udder proper.

The floor of the udder should also be well supported. the medial suspensory ligament divides the udder into right and left halves and holds up the skin along the midline. Proper suspension will have the middle of the floor of the udder rise up toward the body slightly from both sides. Too high a cleft or a flat or dropped floor are both serious faults.

A poorly attached udder becomes pendulous with the milk being carried in a hanging bag swinging between the hindlegs. An udder that is bouncing back and forth as an animal walks is much more likely to develop mastitis than an udder which is carried high and close to the body. Remember, however, that even a well attached udder will appear pendulous when it is not full.

The texture of the udder should be checked by the judge or breeder as he checks the teats. Texture is most accurately judged on an empty udder, so the breeder again has the advantage of being able to most accurately evaluate the animal at home. The judge must palpate the udder in whatever stage of fullness it is presented in the ring. The udder should be soft and pliable and consistent. No bands of tissue should be felt. There should be no lumps or masses. The empty udder should not be a solid meaty structure but should collapse quite a ways. A good functional udder makes lots of milk from only a little mammary tissue.

Teats are readily seen by even the most inexperienced breeder and tend to be grossly overrated when evaluating animals. Only 2-3 out of the total 100 points belong with the teats. The pygmy goat should have two cylindrical teats. They should be balanced in size and placement and should be of proper size to milk with two fingers. Milk should be able to flow freely and there should be only one orifice per teat. Many deviations from the norm are possible and are found in pygmy goats.

Teats that are too small for milking or are too close together for looks and/or comfortable hand milking are a moderate fault. Teats of uneven size or placement are distracting from the symmetry of the udder and are also a moderate fault.

Multiple orifices are an easy straight-forward fault. There are two holes in the end of the teat, both draining one streak canal. This is a serious fault. It will not affect the function of the teat or a kid's natural nursing, but will make hand milking quite messy since milk will flow in two directions at once.

Multiple teats may also be referred to as extra teats or supernumerary teats. Anything over one teat per side is extra. If there is a small, rudimentary teat near the base of, or on the side of a teat, it is called a teat spur. It too is an extra teat. Extra teats may be small or almost the same size as the standard teat. They may be located very close to the teat or way out near the edge of the udder. There may even be more than one extra teat on a side. It is imperative, however, that you do not become overzealous in examining an udder for extra teats. Sometimes a wart or mole or improperly healed cut will cause a lump on the udder and may be suspected as an extra teat. A judge should always give the benefit of any doubt to the animal. If he is not absolutely sure a lump is an extra teat, he should not regard it as such. A breeder should also not be too stringent, but has more lee-way depending on his personal concern about extra teats. Extra teats are a serious fault.

Bifurcal teats may also be called double teats. Two teats (one standard and one extra) are fused along their length, most often with a short split near the tip. The shaft of the teat is wider than normal and is not truly cylindrical in shape. These two fused teats have two streak canals and one orifice for each canal (therefore, two orifices). Each canal may drain separate physical sections of the udder. Occasionally the two canals may not be completely separated internally resulting in a flap-like function of the central wall. This can make milking or nursing difficult. Mastitis would be more likely to occur due to incomplete drainage of all sections of the udder. Bifurcal teats are a disqualifying fault.

The difference between multiple orifices and bifurcal teats is very simple physiologically and anatomically. However, it can be virtually impossible for even an experienced veterinarian to distinguish between the two without proper equipment for examination. This means that a judge should be extremely hesitant to disqualify a doe for double (bifurcal)teats. If there is any doubt (and there almost always will be!), he should give the benefit of the doubt to the doe. The breeder can seek professional assistance in deciding which fault he is dealing with.

Blind teats may have no orifice at all. Obviously no milk can be obtained from that side of the udder. A blind teat may also have an orifice but have no canal inside the teat. This fault can only be determined on a lactating doe. A blind udder can be caused by two blind teats or by other structural abnormalities within the udder. Again, it can only be determined in a lactating animal. All these faults are disqualifying.

Incomplete or improper milking of an animal tends to promote mastitis. A breeder should always check the teats of a newly freshened animal for proper functioning. The kids will do all the follow-up milking and the breeder still has the opportunity to observe any faults. A judge in a showring should not routinely milk the does. If he has strong reason to suspect a blind teat or udder in an animal with a full bag, he should ask the exhibitor for permission to milk first. Then he should wash his hands immediately before stripping some milk from each teat. (The exhibitor should then empty the udder of milk right after the class, or put the doe back with her kids and let them do the work.)

Teat shape is also of importance. Overly large or bulbous teats, or teats that are not distinct from the udder detract from the pleasing appearance of the udder. They may be difficult to milk or nurse and may be more prone to injury.

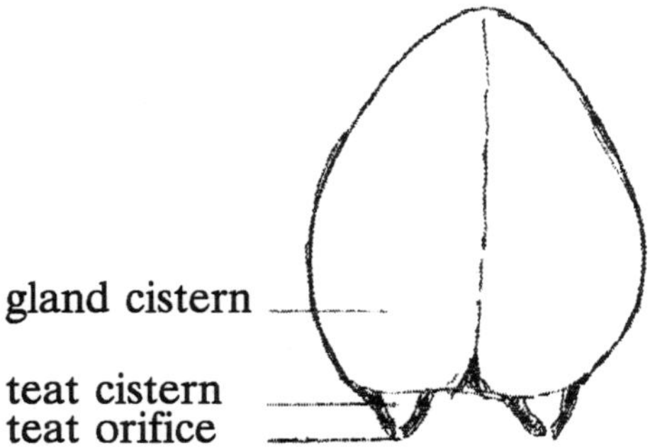

Udder Anatomy

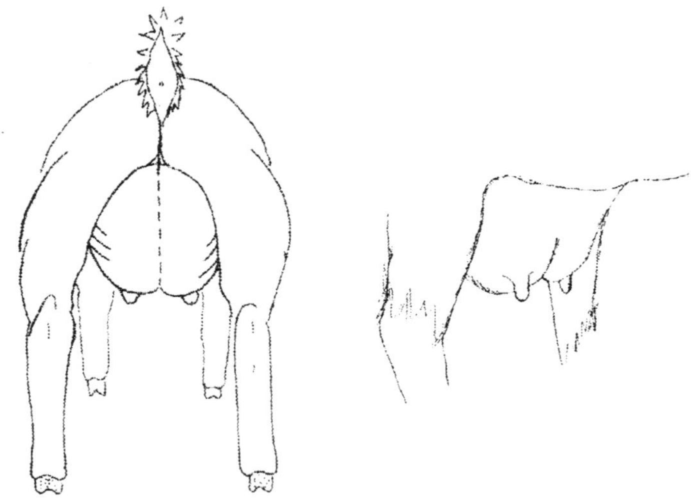

Ideal Udder

Ideal Udder

The udder should be evenly balanced, rounded, well attached, and proportionately sized to the doe. The teats should be symmetrically shaped and placed on the udder. The front attachment should blend smoothly into the abdominal wall in front of the udder. The rear attachment should be wide and high in the escutcheon. The medial suspensory ligament should provide strong support for the floor of the udder.

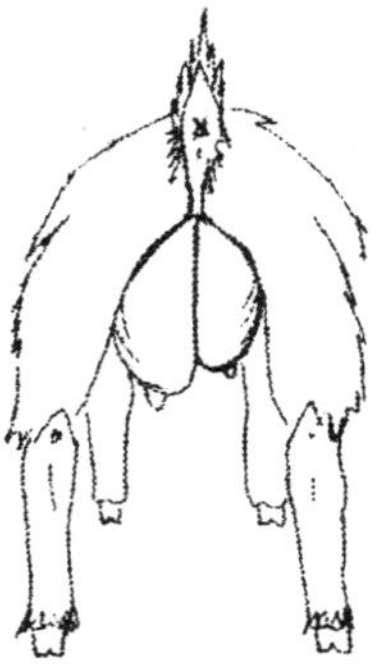

Uneven Udder

The udder should be symmetrical and balanced. An udder may be unbalanced because of conformation, uneven suckling by kids, or disease.

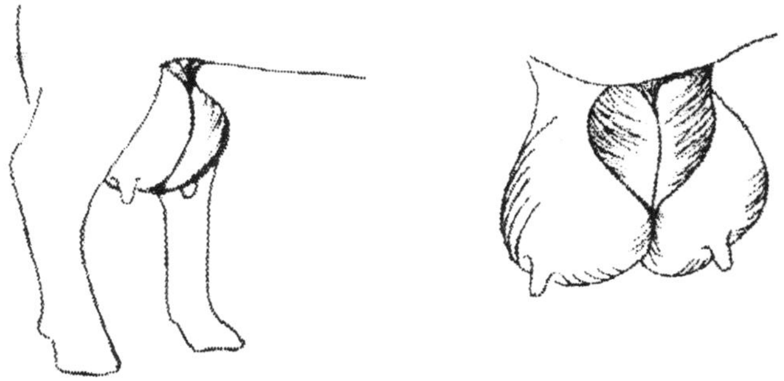

Pocket in Front Attachment

A front attachment that dips back into the udder to attach to the body behind the front of the udder is not desirable. The udder loses support and capacity.

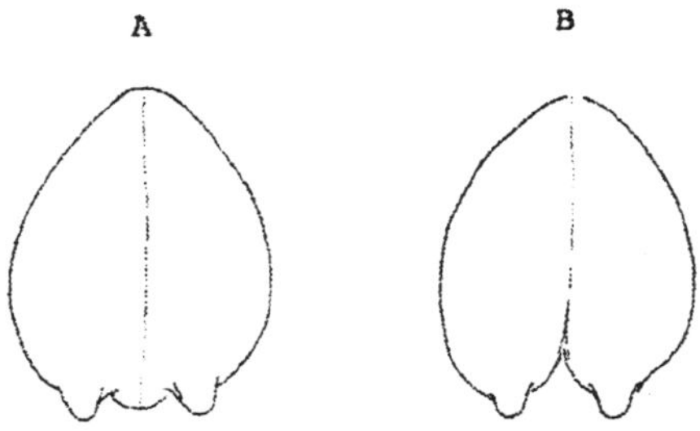

Medial Suspensory Ligament

[A] If the medial suspensory ligament is lacking strength, the floor of the udder will drop and hang down between the teats. [B] An overly tight medial suspensory ligament will cause the udder to appear to be split partially in half.

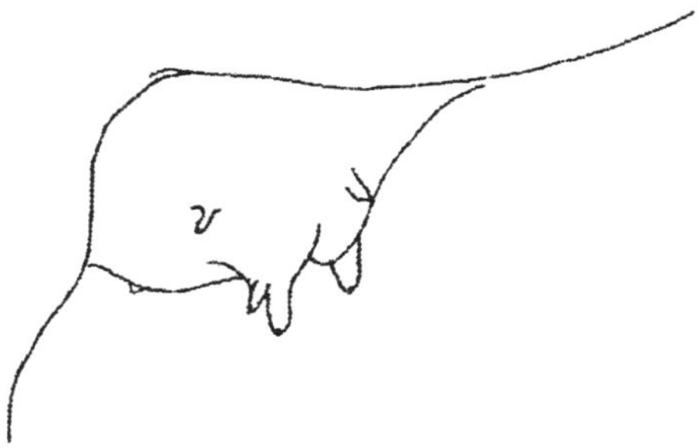

Supernumerary Teat Locations

Each side of the udder should have one teat draining it. Any more than one teat is a supernumerary or extra teat. Extra teats may be located in front of, to the side of, behind, or on the primary teat. Those extra teats that are on the primary teat are also called teat spurs.

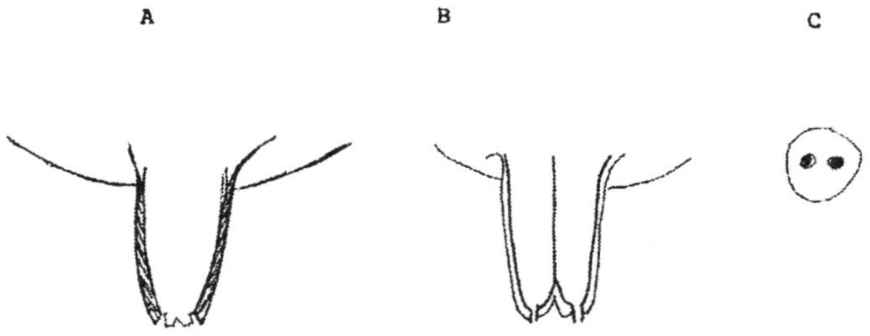

Abnormal Teats

[A] A teat with multiple orifices is usually a normal sized teat with a slightly large, flat tip. It has a single teat cistern but has two separate openings in the tip. [B] A bifurcal, or double, teat is actually two teats that are fused along their length. It will often appear wider than a normal teat and may have a small split into two slightly distinct tips. There are two teat cisterns which may drain the same or separate gland cisterns. [C] The end view of a teat with either of these faults will be the same.

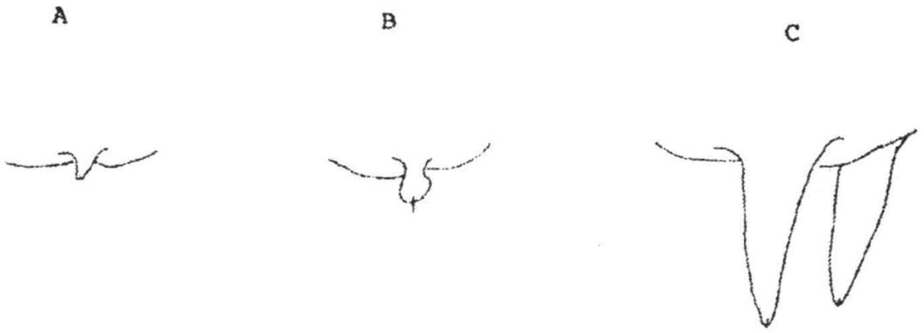

Teat Shape

[A] A teat that is too small will cause a loss of balance and symmetry of the udder. [B,C] Bulbous or conical teats are not pleasing in appearance. None of these faults affect the health or function of the udder. They may make hand milking more difficult but nursing kids will have no complaints.

5

About the Authors:

Author, Lorrie B. Blackburn, graduated from the University of California at Davis, College of Veterinary Medicine in 1968. She has been a practicing veterinarian for eighteen years with a special interest in Caprine medicine. Lorrie has been an active member of the National Pygmy Goat Association since 1977, serving as a Director for seven years and President for three. A licensed NPGA Judge, she set up the original Judges Training Course and acted in the dual role of teacher and chairperson of that committee for four years. She has been a regular contributor to the pages of the NPGA quarterly, "Memo", and her veterinarian columns were an invaluable part of the magazine. She was also on the editorial staff of "Best of Memo". She lives in Orange, California with her husband and two children and the Quillie Acres herd of thirteen Pygmy Goats.

Illustrator, Ann Werts, attained her Bachelor's degree at Chouinard Art Institute, and has been teaching art for nine years in schools in Southern California. Ann was one of the founders of the National Pygmy Goat Association. She helped research and write the Breed Standard and is presently on the Breed Standard Committee. She was a Director of NPGA for many years, served on the Judges Training Committee, was its chairperson for two years, and is an NPGA Judge. Ann lives in Los Angeles County, California with her daughter and Fort Werts herd of Pygmy Goats. She has been raising Pygmies for over nineteen years.

Editor, Lydia Hale, was also Editor of the NPGA "Memo" for three years and Co-Editor of "Best of Memo". She has been raising Pygmy Goats for fourteen years and was one of the founding members of NPGA, serving on the original Certification Committee. She has been a Director for nine years, two years as President. A past member of the Ethics Committee and Judges Training Committee, she is a licensed Judge. Lydia lives with her husband in Pembroke, Massachusetts where they breed their Cottage Farm Pygmy Goats and Cotswold sheep.

All Publishing Company
10951 Meads Avenue, Box G
Orange, CA 92669

ORDER FORM

Name:______________________________

Address:____________________________

City:______________State:____Zip:________

Please send:

______ copies of The Illustrated Standard of the Pygmy Goat

______ copies of Pot-Bellied Pet Pigs, Mini-Pig Care and Trainning

Total for book(s) @$9.95 each	
Sales tax @6.25% (Calif. only)	
Shipping and handling	
Total amount enclosed	

Shipping is $2.00 for the first book and $1.00 for each additional copy. Book rate may take 4 to 6 weeks. First class shipping (and non-U.S.) is $4.00. Visa and MasterCard accepted. Prices are subject to change without notice. Please write for quantity discount rates.

All Publishing Company
10951 Meads Avenue, Box G
Orange, CA 92669

ORDER FORM

Name:________________________________

Address:______________________________

City:______________State:____Zip:________

Please send:

______ copies of The Illustrated Standard of the Pygmy Goat

______ copies of Pot-Bellied Pet Pigs, Mini-Pig Care and Trainning

Total for book(s) @$9.95 each	
Sales tax @6.25% (Calif. only)	
Shipping and handling	
Total amount enclosed	

Shipping is $2.00 for the first book and $1.00 for each additional copy. Book rate may take 4 to 6 weeks. First class shipping (and non-U.S.) is $4.00. Visa and MasterCard accepted. Prices are subject to change without notice. Please write for quantity discount rates.